ABRAHAM'S
CHILDREN

ABRAHAM'S CHILDREN

Race, Identity, and the DNA of the Chosen People

JON ENTINE

GRAND CENTRAL
PUBLISHING

New York   Boston

Grand Central Publishing
Hachette Book Group USA
237 Park Avenue
New York, NY 10017

Visit our Web site at www.HachetteBookGroupUSA.com.

Printed in the United States of America

First Edition: October 2007
10 9 8 7 6 5 4 3 2 1

Grand Central Publishing is a division of Hachette Book Group USA, Inc.
The Grand Central Publishing name and logo is a trademark of Hachette Book Group USA, Inc.

Library of Congress Cataloging-in-Publication Data

Entine, Jon.
  Abraham's children : race, identity, and the DNA of the chosen people / Jon Entine. — 1st ed.
    p. cm.
  Includes bibliographical references and index.
  ISBN-13: 978-0-446-58063-2
  ISBN-10: 0-446-58063-5
  1. Jews—Identity. 2. Jews—Origin. 3. Jews—History. 4. Physical anthropology.
  5. Human genetics. 6. Race. I. Title.
  DS143.E58 2007
  305.892'4—dc22

                                                                2007006026

Book design by Giorgetta Bell McRee

To Madeleine and her special friends—
Joey, Langley, and the three boys—who bring joy
into our home each and every day.

CONTENTS

ACKNOWLEDGMENTS AND AUTHOR'S NOTE

I am especially grateful for the backing of the American Enterprise Institute for Public Policy Research and its president, Chris De-Muth, who saw the value in this book when it was just a glimmer of an idea. Particular thanks goes to AEI's National Research Initiative, which generously provided support for the research and writing, and most especially to NRI's first director, Kim Dennis, who has become a wonderful friend, and to Henry Olsen, who succeeded Kim.

My agent, Glen Hartley, was always patient and encouraging, and my lawyer, Leon Friedman, was instrumental in keeping this project afloat when it appeared it might not survive the politics of publishing. I am very lucky that the book was placed in such able hands at Grand Central Publishing. Vice president and long-time friend Rick Wolff; my editor, Les Pockell, who made this book much better than when it arrived on his desk; assistant editor Celia Johnson; Elly Weisenberg, my indefatigable publicist; and the entire staff at Grand Central Publishing have provided invaluable guidance and support at every turn.

Many, many people devoted endless hours helping in the research and the critiquing of manuscripts. Donna Lyons, who did double duty as researcher and my daughter's nanny, adeptly ferreted out hundreds of DNA studies. Kevin Brook, who is an expert on Khazarian Jews, reviewed the book in its many stages. Bennett Greenspan of Family Tree DNA, a genetic genealogy testing service, provided invaluable contacts, data, and advice. Mary-Claire King, Karl Skorecki, and John Derbyshire critiqued significant portions of the manuscript. Others who reviewed sections at various stages of writing or provided key research include Gregory Cochran, Henry Harpending, Stan Hordes, Michael Hammer, Sergio Della-Pergola, Doron Behar, Batsheva Bonné-Tamir, Joel Zlotogora, Neil Bradman, Mark Thomas, Charles Murray, David Goldstein, Tudor Parfitt, Vincent Sarich, Moein Kana'an, Ya'akov Kleiman, Benyamim Tsedaka, Bruno Mauer, and Father William Sánchez.

Although *Abraham's Children* draws on many historical sources, I would like to note a few books that I found particularly helpful. The six-volume reference collection *The Anchor Bible Dictionary* (New York: Doubleday, 1992) provided historical and interpretative context for the biblical narrative. Paul Johnson, *A History of the Jews* (New York: HarperCollins, 1987); Raymond P. Scheindlin, *A Short History of the Jewish People* (New York: Oxford University Press, 1998); and Eli Barnavi (ed), *A Historical Atlas of the Jewish People* (New York: Schocken, 1992), were well consulted. Rabbi Joseph Telushkin, *Jewish Literacy* (New York: William Morrow, 2001), and Bruce Feiler, *Walking the Bible* (William Morrow, 2001), were delightful and informative. For interpreting genetic anthropology, no book was more essential than Luca Cavalli-Sforza, Paolo Menozzi, and Alberto Piazza, *The History and Geography of Human Genes* (Princeton: Princeton University Press, 1994), one of the most fascinating and important books of the twentieth century.

Because of the breadth of *Abraham's Children*, it inescapably touches many disciplines, addressing wide-ranging issues in science, history, and religion, including Christianity and Islam, about which I am far from being an expert. For that reason, some readers may find aspects of this book either too simple or too complex, and for that I apologize in advance.

The Bible, both the Hebrew original and the New Testament, is the most widely translated book in the world. You may find that your favorite passages are rendered slightly differently in this book. For the sake of consistency, I chose two English translations as the sources for all the biblical quotations. The citations from the Hebrew Bible come from the original translation of the Masoretic—the traditional Hebrew text found in *Tanakh: The Holy Scriptures* (Philadelphia: Jewish Publication Society, 1985). Quotations from the New Testament are from *The New Oxford Annotated Bible*, 3rd Edition (New York: Oxford University Press, 1989).

Readers will note that in keeping with the custom in scientific publications, *Abraham's Children* uses the nonsectarian epochal term BCE (Before the Common Era) instead of BC and CE (Common Era) instead of AD.

PART I

IDENTITY

CHAPTER 1

THE DEAD SEA SCROLLS OF DNA

Moses called the blistering sands of ancient Canaan "a great and terrible wilderness." On a cloudless day, on a plane at 10,000 feet, the allure is immediately apparent. The vast plateau of the Negev, the northern Sinai in Egypt, and Jordan's southern desert rarely looks like the sweeping, endless dunes of popular imagination. More commonly, it resembles a visually dissonant landscape of dusty hills lined with scarps and dotted with scraggly bushes and rocks. It is untamed beauty, with barren stretches interrupted only by Bedouin tent villages and archaeological oases that seem to come and go with the wind.

My throat clogged in the heat of the afternoon desert air as I walked off the small Arkia jet and onto the melting tarmac in Eilat, Israel's Red Sea beach town at the tip of the Negev Desert. Set in a stunning location along the Gulf of Aqaba, which divides the Sinai from the Arabian Peninsula, this palm-fringed city had thrived by attracting sun-worshippers from France, Italy, and Scandinavia. Scuba divers still explore the deep aqua waters, ablaze with coral reefs, that wash on its shores. But like the rest of Israel, this once-gleaming resort is struggling, another casualty of the tension that engulfs the Middle East. My beachfront hotel stood empty, save for

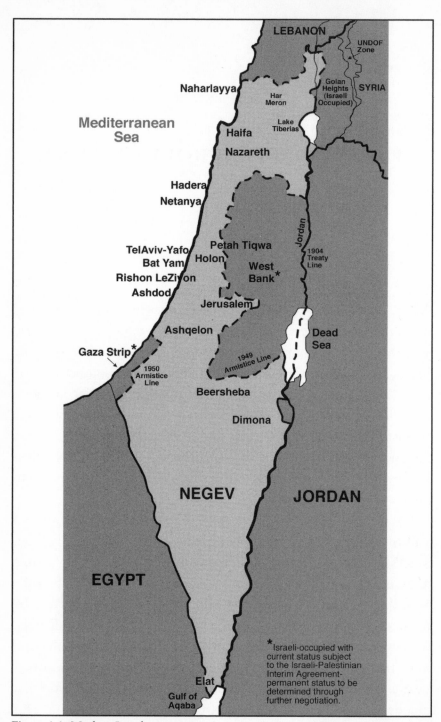

Figure 1.1. Modern Israel.

a large group of unruly students from suburban Tel Aviv vacation-
ing on the cheap.

Israel is central to Western culture and religion despite—or perhaps because of—its challenging topography, where the identity
of the Israelites was forged. This scruffy region is a "land flowing
with milk and honey," as God promised to Moses as he began his
desert sojourn in Canaan, but mostly in the spiritual sense, for life
here has always been a test. The desert did not share in the prosperity spilled along the fertile banks of the bountiful rivers that
spurred the rise of two other centers of ancient civilization, along
the Nile in Egypt and the Tigris and Euphrates in Mesopotamia.
The "mighty" Jordan River that divides Israel and Jordan—ancient
Judea* and Samaria to the west and Edom to the east—was always
richer in folklore than volume. Today in many spots it is little more
than a trickle.

My visit to Israel was a deeply personal journey, spurred by the
tragedy that DNA has visited upon my family. Although raised as
a Reform Jew, dutifully bar mitzvahed and confirmed, and even
though I had majored in religion and philosophy in college, I had
long since lost my faith. The final break came in my teen years, in
1970, when my mother died of ovarian cancer, an unrelenting disease that spread to her brain. She had fought courageously for years,
undergoing chemotherapy that left her bloated and caused her hair
to drop out. My father, a dentist turned drug sleuth, scoured the
world for potential treatments, eventually finding a promising new
drug in Israel, unapproved at that time in the United States, that
he smuggled home in his luggage in a desperate attempt to keep
her with us a little longer. Her doomed struggle has left me with a
haunting memory. Although I was only fifteen when she began the
secret therapy, she asked me to give her the daily injection that we
all viewed as her long-shot lifeline; she believed I would administer

---

* Judea is the Greek name adapted from "Judah," which was the original name of the
southern kingdom of Israel, derived from the tribal name "Judah."

the shot more gently than my dad. I watched her every day, too closely, as the cancer spread.

My mother was neither the first nor the last person in my family to suffer from cancer. Two years before, my grandmother, living in our home, passed away from breast cancer. It also took my mother's younger sister, in her thirties. Because science was not yet able to explain the origins of this type of cancer, our family believed the three deaths were tragically unfortunate coincidences. As we were later to learn, my mother and her family were victims not of bad luck but of a bad gene.

A few years ago, I received a horrific call from my older sister, Judy. She too had been diagnosed with breast cancer, a rapidly spreading kind similar to the version that had savaged our aunt, grandmother, and mother. Thirty years after my mother died, the science of genetics had come a long way. Judy underwent a genetic test that revealed that her cancer was almost certainly caused by a mutation on one of her genes. Identified just over a decade ago, it was dubbed the BRCA2, or *BReast CAncer* 2 mutation 6174delT, one of three breast cancer mutations that are particularly common among Jews. My family tree disappears into the nineteenth-century eastern European diaspora, so it's difficult to trace the history of this wayward gene in my maternal line. The only thing that can be said with near certainty is that it's a tragic marker of our family's Jewish ancestry.

It's estimated that one in forty-three Jews (about 2.5 percent), women and men, carry one of these three gene faults. Being a carrier doesn't mean you have or will definitely get breast or ovarian cancer. It does mean that cancer is much more likely to develop because your cells are one critical step further along the road to becoming cancerous than they would be if you didn't carry the mutation. While the general population faces a 10 percent risk of developing breast cancer, the risk for women with one of these mutations may rise during their lifetime to more than 80 percent, even for women with no family history of the disorder. For many female carriers, it used to be an almost certain death sentence.

Defying the grim odds, Judy fought the disease to a standstill, showing remarkable courage; tapping into the support of her husband, family, and friends; and seeking the advice of medical experts and geneticists who were not available to my mother decades ago. She's now healthy and disease-free. When she was diagnosed, she had urged me and Joan, the middle child in our family, to get tested. Like many people who learn they might be carrying a genetic defect, Joan didn't want the burden of knowing if she was a carrier. She had no daughter to whom she could have passed along the mutation, so she declined to be tested.

I chose differently. By pure coincidence, I had written about the DNA revolution in previous books on genetic engineering in agriculture and on the role genes play in influencing which races do best in which sports.* But now genetics was a family and deeply personal concern. I have a young daughter who might have inherited the disease mutation from me, so I decided to be tested.

The news was not good. I too carry the cancer mutation. Its effects on men include a slight risk of breast cancer—yes, I now get my breasts squeezed and poked each year—and slightly increased risks of pancreatic cancer, prostate cancer, and melanoma of the eye. Although I tested positive, I was not allowed to go the next step and find out whether my daughter had inherited this potentially killer gene from me. Myriad Genetics, which isolated the BRCA mutations, does not allow its test to be administered to minors. It holds that a mutation in someone so young would not yet be showing its effects, and as there are no preventive measures, a positive test would needlessly frighten. The decision was made for me. We would have to wait until she was eighteen, and then it would be her choice whether to proceed with the test.

We've all heard the phrase "We are, regardless of race, 99.9 percent the same." Yet here was a pinpoint of DNA that suggests that

---

* *Let Them Eat Precaution: How Politics Is Undermining the Genetic Revolution in Agriculture* (Washington, AEI Press, 2006); *Taboo: Why Black Athletes Dominate Sports and Why We're Afraid to Talk about It* (New York: PublicAffairs, 2000).

maybe human population groups are not quite so alike as the common wisdom holds. However slight the genetic differences (and geneticists now believe that they are far greater than the 0.1 percent that previously had been estimated), they are defining. They contain the map of my family tree back to the first modern humans. They catalog my extended family's vulnerability to many diseases. And they mark me indelibly as a Jew. Almost every minority group, at one time or another, has faced being branded based on a superficial understanding of how genes work. The inclination by politicians, educators, and even some scientists to highlight the threads that bind together the world's diverse populations and underplay our separateness is certainly understandable. But it's also misleading. DNA ensures that we differ not only as individuals but as groups.

The modern age of genetics is symbolized by the famous image of the twisted ladder of chemical crossbars of deoxyribonucleic acid. It took years after the basic structure of DNA was identified before scientists could draw even a rough outline of the script that nature took billions of years to compose. "Today we are learning the language in which God created life," noted President Bill Clinton in June 2000, invoking religious metaphors in announcing the first working draft of the human genome. Francis Collins, director of the National Human Genome Research Institute, who was an atheist and is now a practicing Christian, drew an even more direct religious reference at that news conference. "We have caught the first glimpses of our instruction book, previously known only to God," he said.

This book offers a story of faith and science. I embarked on this journey as a skeptic by nature and profession, and from my vantage point as a "High Holy Day Jew"—what many Orthodox believers, without humor, call a "gentile." But this book suggests that religious identity extends beyond beliefs. Our genes carry meaning. This ancient script now being deciphered is literally lifting the curtain on God or Nature's plan. While often at odds, religion and science are spinning an interlaced narrative of identity. It's a story that begins in the ancient Middle East.

ISRAELITE GENES

What can DNA tell us about the shared legacy of the Israelites?

Christians, Muslims, and Jews have always recognized a common spiritual heritage that originates with "Abraham, our father"— the progenitor of 2 billion Christians, 1.3 billion Muslims, and 13 million Jews. "I will establish My covenant between Me and you," God announces in Genesis to Abraham, promising that his descendants will be "as numerous as the stars of the heaven and the sands on the seashore. . . . I assign the land you sojourn in to you and your offspring to come, all the land of Canaan, as an everlasting holding. I will be their God." Abraham and his descendants will forever be unique, their future in the Promised Land irrevocably bound to the grace and favor of God.

What of this story is true? What evidence exists to support the central narratives of the Hebrew Bible—the birth of the Israelite people under Abraham, the Exodus, the golden age of the Davidic Empire, and the story of the Lost Tribes, among others? After all, no existing records other than the Hebrew Bible refer to Abraham, a sizable Israelite presence in Egypt, or even the Exodus.

Many early religious leaders, Jewish and Christian, including Saint Augustine in *De Genesi ad litteram*, written near the end of the fourth century CE, professed that aspects of the Hebrew Bible should not be taken at face value. But Christian orthodoxy ultimately overwhelmed theological rationalism. From the fall of the Roman Empire until the mid-nineteenth century, when Charles Darwin proposed a heretical naturalistic explanation of human origins, a literal interpretation prevailed. Over the past century, skeptics and scholars known as biblical minimalists have chipped away at many key stories. Nature's cemeteries have yielded skeletal remains, coins, pottery shards, and flint tools with exotic engravings that have enriched and often rewritten classical narratives. But like the ancient biblical accounts, the bones and stones of history have left us with chapters abbreviated and characters missing. The metamorphosis of the scattered Israelite tribes into a proud and irascible people has remained only dimly visible.

Questions remain. Were Abraham, Moses, and David real people? What happened to the Twelve Tribes? Can some modern Jews actually trace their ancestry as Jewish priests to Aaron? Did descendants of King Solomon and the queen of Sheba build the fabulous stone palace in the heart of Africa known as the Great Zimbabwe? Is Britain or the United States the "New Jerusalem" foretold in the Bible? Are the American Indians descendants of Abraham, as some Mormons believe? What happened to the Jews who converted to Christianity during the Spanish Inquisition? Who are the real "chosen people"—Arabs or Jews? Are the majority of today's Jews descendants of non-Jewish Eurasians and Europeans? If the maternal lineage of many modern Jews begins with gentiles, as history and DNA suggest, and they did not formally convert to Judaism, what determines "Jewishness"? If blood ties were paramount to our Israelite forebears, then our genes, protected against the ravages of time and passed on from generation to generation, must carry some record of the Israelites, the origins of Christianity and Islam, and the long and contentious travail of the Jews.

DNA has for good reason been described as "the book of life." The cells in the human body make up a vast library that carries the story of all life on earth, which began with our protozoan ancestors, the organisms that lived some 4 billion years ago. Each of those 100 trillion cells is a microscopic, soupy, fluid-filled bag with a tiny blob called a nucleus floating inside. That's where our life's book is archived. Each nucleus contains forty-six chromosomes—two pairs of twenty-three spiral DNA strands. They hold the code for how each cell should work and therefore how traits—from the shape of our nose to our temperament—shall be expressed.

The portions of the DNA that contain the recipe of life are known as genes. The "sentences" for these instructions are proteins, which are made up of amino acids, or "words," each of which consists of "letters"—nucleotides. There are enough letters in the human genome—more than 6 billion nucleotides that make up 3 billion base pairs—to fill thousands of Bibles. This immense encyclopedia of evolution fits inside the nucleus of every cell. If stretched out, the DNA in each human chromosome would extend an inch or more, but it is compacted into a microscopic bolus of only 0.001 inch.

Each cell has a full set of DNA instructions. So, for instance, a hair cell also has the genetic information on how to make teeth or eyes, but those sections of the DNA are shut down in that particular cell. It's why crime scene investigators can take a sample of DNA from any part of the body and get a distinct signature of that individual and how genetic identity seekers can map the story of our ancestral history back to biblical times. Using unique DNA markers, our female ancestral lineage can be traced through the energy center of cells, known as mitochondria. The narrative of the male line is embedded on the male sex chromosome, Y. Yet another set of ancestral stories emerges from the vast majority of genes, known as autosomes, found on the rest of the human genome.

What our DNA reveals is at once comforting and disquieting, for it confirms our common humanity and yet challenges deeply held beliefs of equality. We often hear talk in our cosmopolitan capitals that we live in a global village and a world melting pot, but our genes tell a more complex story of diversity and differences. Over many thousands of years, twigs off the human family tree—bands of extended relatives, clans, tribes, and what scientists call populations—migrated throughout the world. Although "the urge to merge with a splurge," as Cole Porter once put it, has inexorably erased many of the most visible signs of our ancestral roots, we are still a long way from shedding the heritage embedded in our genes.

The story contained in our DNA raises the taboo issues of race, disease, and intelligence. Genetic anthropology and genealogy are tightly bound to the worldwide effort to address many behavioral and medical problems, which is a key impetus behind the Human Genome Project. Genetic differences, and sometimes just one gene, can confer near-certain death sentences, while other people are mysteriously spared. There are thousands of genetic disorders, hundreds of which disproportionately affect one racial or ethnic group.

Although rare in blacks and Asians, cystic fibrosis is a common lethal genetic disease in those of northern European ancestry. Whites are more likely to get multiple sclerosis than all other population groups, while blacks and some Mediterranean populations are susceptible to sickle cell anemia. Some 2 million whites worldwide now carry copies of a mutant gene that makes them immune to human immunodeficiency virus (HIV) infection. Rare mutations may help insulate some southern Asians from severe acute respiratory syndrome (SARS). The presence of one gene is a potent risk factor of Alzheimer's for whites, but not for blacks. The variant of one gene may explain why black women have twice the risk of premature delivery than women of European ancestry. Those of African ancestry are more susceptible to heart disease and are 50 percent more likely than whites to die of colorectal cancer, even if they receive the same treatment. Irish and others of Celtic ancestry are disproportionately victimized by Dupuytren's disease, also known as claw hand. One mutation accounts for the sensitivity of the Japanese to alcohol, while another gene variant carried by at least a fifth of all Semites helps them break down liquor in the bloodstream, protecting them against alcoholism.

Although geneticists generally avoid using the term "racial" to characterize differences that show up more in one population than others, ancestry matters. Because modern humans move around and fool around far more expeditiously than their ancient ancestors, modern "races" and ethnic groups are fuzzy at the edges and overlapping. As a rule, the more historically isolated a population—because of geographical or cultural barriers—the more distinct its genetic makeup. Tantalizing clues about the origin of diseases have prompted researchers to focus on a number of geographically circumscribed groups; Icelanders, Finns, indigenous Ainu of Japan, American Indians, Costa Ricans, Maori of New Zealand, Sardinians, Basques, rural Chinese, and various West African tribes and their descendants are examples. Known as founder populations because they tend to preserve the genetic makeup of the founders, these groups have two key attributes: only a few thousand primary ancestors and little intermarriage in succeeding generations. Other genetic islands have been shaped by strong religious and cultural beliefs—Gypsies, the kindred clans of Mennonites and Amish, Hutterites, and Parsis. But no gene pool has been more crucial for DNA research and the quest to develop medical cures than the diaspora communities of the Jews.

Jews have unique advantages as candidates for genetic study. Since being expelled from biblical Palestine to the far corners of the globe, they have congregated in many tightly knit but intricately linked communities. For the most part, they were endogamous—they rarely married outside the religion, at least until the twentieth century. That fidelity has proved a gold mine for DNA researchers. "Jewish genetics," as it has been called, is at the center of a worldwide quest to solve the puzzle of disease and unlock the backstories of humanity. Archival vaults once thought buried in time are being pried open. Provocative answers to questions of ancestry and identity, once considered answerable by faith alone, might be found in our DNA. It's the genetic equivalent of history's Dead Sea Scrolls. Consider the story of Father William Sánchez.

CHAPTER 2

IDENTITY COMPLEX

Let us remember all of our Jewish brothers and sisters, who through their faithfulness to One God, inspired all Christians."

Resplendent in a red and white robe on this Palm Sunday Eve, Father William Sánchez solemnly marks what he calls the "Passover of Jesus," the journey of the king of the Jews that begins with his entry into Jerusalem.

"Praise him, descendant of Abraham, Isaac, and Jacob, give glory to him, revere him and all you descendants of Israel," he intones in a voice that reminds one of Placido Domingo, the Spanish tenor whom he resembles in looks and presence.

While Jews around the world mark the Lord's decision to spare the Hebrews and preserve the Western tradition of monotheism, the Mexican American congregants at Saint Edwin's, a modest church carved out of an aging parish hall in south Albuquerque, New Mexico, honor the Passover sacrifice through the prism of Jesus's crucifixion and resurrection. Father Bill, as his parishioners affectionately call him, lifts the goblet filled with the sacramental wine to his lips and gives voice to Jesus's words as he took his place at the Passover table with his disciples.

"Take this cup and share it among yourselves. This cup is the new covenant in my blood, which will be shed for you and for all so that sin will be forgiven."

The invocation of a blood ritual is not an idle one. From this moment in biblical time onward, for followers of Jesus, the covenant would be grounded not in blood ancestry but in the symbolism of blood represented by faith. This defining event marked a fundamental divide that would forever separate the religious identity of Jews from the followers of the two major religions that Judaism would spawn, Christianity and Islam.

Or would it?

"Christianity may have changed the blood covenant from literal to spiritual, but it does not deny the ancestral connection of Jews to Christianity," said Father Bill later that evening in his spartan office in the rectory building across from the church. "And the fact is," he said, his eyes looking at me with a gaze both weary and hard, "I'm Jewish. I believe in the word and message of Jesus, but I am a Jew as well."

Father Bill is in his midfifties, but the harsh reaction of the Catholic hierarchy to his spiritual journey has made him feel much older. He rested his hand gently on the 6-inch-high stack of DNA tests piled on his desk documenting his own ancestry. "For me, as a Roman Catholic, we're very Jewish in our traditions, in the liturgy," he said. "God chose the Israelites. Jesus was Jewish. He died a Jew. I'm just acknowledging that fact, that spiritual fact, within myself. But now it has a literal reality as well. It's embedded in my genes, in my DNA."

These were startling and perhaps reckless words for an idealistic parish priest who believes that sharing his belief in his Jewish ancestry is part of his calling in the name of Christ. An alphabet soup of genetic data has challenged his faith and threatens to fracture his relationship with Catholic authorities. He knows that every word of every interview he gives, every conversation with a journalist, and every public speech will be reviewed by Catholic officials who are irate with him because of his outspokenness in discussing his "Jewish roots." "God brought my ancestors to New Mexico from

Spain for a reason," he said, looking determined, but in the soft-
ness of his voice, sounding almost stricken. "I'm very comfortable
knowing that I have ancestral roots that deepen the shared religious traditions of Christians and Jews"—his voice trails off in a
whisper—"even if the church is not."

WHO IS A JEW?

Father Sánchez's spiritual quest is a familiar one for Jews, who
have spent their history wrestling with their identity. "The remnant
of Jacob shall be in the midst of many peoples, like dew from the
Lord, like droplets on grass," reads the biblical prophecy in Micah
5:6, which was indeed prophetic. Jews make up a tiny 0.25 percent
of the global population, but pockets of adherents are found in
almost every culture. And the genetic seeds of the Israelites can be
found in many ethnic groups and religions. The explosion of interest in genetics has revived one of Western society's longest-running
debates: who is a Jew? Identity may turn on the mysteries encoded
in a lock of hair or a single drop of blood. Can we find an answer
in faith or ethnicity, scripture or ancestry?

Jewishness is an amalgam of characteristics. Many Jews who
do not practice their religion consider themselves Jewish because
of family heritage or because they identify with the history of the
Jewish people. Forged over centuries of forced and self-chosen
isolation and a fierce resistance to assimilation, Jewish identity is
suffused with biblical lore and nested like a Russian wooden *ma-
tryoshka* doll in layers of nationality, race, and culture. Ancestry is
the mortar that has sealed the meaning of Jewishness.

By the first century CE, the great Jewish historian Flavius Josephus identified himself in *The War of the Jews* with the words,
"I Josephus, Son of Matthias, a Hebrew by race . . ." His motiva-
tion was certainly political as well as theological: he was intent
on rationalizing Rome's crushing victory over the last vestige of
a nation of Jews, preserving Jewish dignity, and perhaps rescu-

ing Judaism from desecration and oblivion. Fast-forward to nine-teenth-century Europe, and Jewish racial identity remains intact. Although the British prime minister Benjamin Disraeli was bap-tized a Christian because his father, Isaac, left the faith after a quarrel with the elders of his synagogue, he never ceased cham-pioning the "Hebrew race" that he was born into. "All is race, there is no other truth," he wrote. Disraeli called race the "key to history" regardless of "language and religion [for] there is only one thing that makes a race and that is blood" and there is only one rule, "the aristocracy of nature." For Disraeli, as for most Europeans of Jewish ancestry, lineage trumped faith. "Believing in his own chosenness without believing in Him who chooses," Hannah Arendt would later call it.

The persistence of Jewish distinctiveness is all the more remark-able because Judaism has been a religion on the run for most of its history. Jewish identity managed to survive only because it has continued to embrace the notion of tribal identity, rather than de-fining itself strictly by doctrinal beliefs. In biblical times, religion and tribal identity were tightly bound. The Egyptians, Assyrians, Babylonians, Philistines, Greeks, and Hittites all had their distinct pagan beliefs and lived, in oscillating competition and cooperation, with the Hebrews of Canaan. Only the religion of the Israelites has survived in this form. Judaism is distinct among the great religions of the West in retaining descent side by side with belief as defining components. "Why are the Jews like the fruit of the olive tree?" asks a famous midrash, a rabbinical interpretation of ancient texts. "Because as all liquids mix with each other, but the oil of the olive does not, so Israel does not mix with the Gentiles. As the olive does not yield its oil unless it is crushed, so Israel does not return to God unless it is crushed by affliction."

Chosenness has fed intense feelings of admiration but also pe-riodic waves of curiosity, wariness, and hostility by the rest of the world. Separateness came to be seen as a self-imposed state that carried within it sometimes horrific consequences. In the tortured view of Franz Kafka, there was "no exit" for Jews from their historical predicament: the enslavement in Egypt under Pharaoh Rameses II; the pursuit by the Amalekites in the desert on the journey to the Promised Land; the two biblical exiles; the destruction of the Second Temple and the forced diaspora; the medieval Crusades; the Spanish Inquisition; and the ultranationalism and racism that would consume Europe—all inseparable links in the chain of Jewish history and identity. Even the rush to assimilation that has marked the Jewish experience in the nineteenth and twentieth centuries is quintessentially Jewish, for nothing can be more Jewish than not wanting to be. For Jews, ancestry is destiny. The more one tries to abandon his or her Jewish roots, the more Jewish he or she becomes.

The Jewish tradition sometimes referred to as "exceptionalism," relentlessly reaffirmed throughout the centuries in scattered communities of Jews around the globe, has left a deep genetic footprint. The key to survival, Jews reckoned, was fidelity—to their religion and their spouse. It was a brilliant strategy for a minority community and would evolve into a unique formula of ethnocentrism that would be emulated in exile for millennia to come. Until the last two generations, Jews were noted for what scientists call consanguinity and endogamy. They tended to marry within their community, even in eras when secularization and assimilation periodically drained the centrality of God and the messianic message. Conversion, while allowed, was not so simple as accepting God; it required a demanding period of study and reflection, which greatly limited the number of converts. Halakhah, Jewish law, further holds that rabbis should make three vigorous attempts to dissuade a person who wants to become a Jew. For converts who traveled the righteous gauntlet, their Jewishness was thereafter passed on by blood.

Over many centuries, forced and voluntary conversions to Christianity, pogroms, and massacres pruned branches off world Jewry, but faithful Jews turned fate to their favor. Isolation and dispersion served as an ironic survival mechanism, transforming the demographics of Jewry in the process. In the early medieval period, the world population of Jews was below 1 million and shrinking, with almost all of them in Muslim lands, with southern Spain the capital. Sizable numbers of Jews found themselves caught between Islam and a resurgent Christianity, a vise that would tighten with the Inquisition.

At first, the Jews of Europe did not fare well. They dwindled in number to approximately 25,000 in the thirteenth and fourteenth centuries. Through this Jewish Dark Ages, religious Jews convinced of the chosenness of their people kept Judaism from disappearing. Although Jews were a tiny minority in the villages of central and eastern Europe, their commitment to literacy made them invaluable as tradesmen. More prosperous than most, the Jewish population began to steadily increase. The baby boomlet was stalled by the Cossack massacres in Poland and the Ukraine in the mid-seventeenth century, but it commenced again in the 1700s. By the 1765 census, the number of European Jews stood at about 800,000. The Enlightenment then unleashed an era of unheard-of prosperity that lifted even the tiny boats of European Jewry. For the first time, European Jewry eclipsed in number the shriveling communities of Middle Eastern and Spanish Jews, the result of the vastly better living conditions in Europe compared to the Muslim world. By the fin de siècle, the Jewish population had soared past 10 million, eventually peaking at nearly 17 million before World War II.

Then came the Holocaust.

While Jews during the Nazi era faced liquidation by genocide, today they face dissolution by conversion, assimilation, and indifference. Approximately 50 percent of Jews in Europe and North America (80 percent of them Ashkenazim) marry non-Jews. Geneticists call this demic diffusion—the spread of genes from a smaller population into a larger one. This demographic trend began in nineteenth-century Europe, when Jews first entered secular, middle-class society in great numbers. It accelerated as refugees flooded into the United States, which became the new center of world Jewry and where religious taboos were fewer and assimilation the norm. In the 1920s, fewer than one in one hundred Jews in the United

States married non-Jews. But while millions may have lived in Jewish neighborhoods, been considered Jewish by the larger Christian community, and even married other ethnic Jews, their commitment to the religion was waning. Assimilation softened the taboos on mixed marriages. Within decades, the number of Jews marrying gentiles and those leaving the faith altogether would eclipse the population of Jews in America, a pattern repeated in France, England, and other Western capitals with large Jewish populations.

The number of Jews worldwide is thought to be about 13 million, although estimates are invariably hazy because of the complex notion of Jewish identity. The total has been dropping by 300,000 each year. Approximately 10 million Jews are Ashkenazim, a word derived from the Hebrew word for "German," which suggests their recent European roots. (About 90 percent of the approximately 5.5 million Jews in the United States are Ashkenazim.) The remainder are mostly Sephardim (from the Hebrew word "Sefarad," meaning "Spain"), who trace their ancestry to Iberia or North Africa, the center of diaspora Jewry until the medieval inquisitions; and indigenous Jews from the Middle East known as Orientals, most of whom now live in Israel.

Endemic intermarriage and a low birthrate will continue to slice into the marrow of Judaism. With the Jewish fertility rate below the level necessary to replace the current population, in just a few generations, two out of every three Jews could disappear. While disastrous for the long-term future of the religion, this seemingly irreversible trend has proved a boon for geneticists and genealogists. Most Jews who intermarry raise their children faithless or as gentiles, in effect removing them from the "Jewish gene pool." Those who assimilate out of the religion reinforce the genetic distinctiveness of faithful Jews. The diaspora history of the "wandering Jew" makes it impossible to draw up a simple genetic profile of Jewish identity. But the DNA pruning process has preserved an ethnic core population with a common ancestry that many Jews believe defines Jewishness. The genetic legacy of the ancient Israelites is also preserved in millions of unbelievers, Christians, and Muslims destined to carry their biblical inheritance forever in their genes.

JEWISH PRIESTS

Father Bill's search for a Jewish identity he did not even know he had lost began one evening just after Easter in 2001. After a grueling week of church responsibilities, he was relaxing at his parents' home in Santa Fe when he happened upon a public television special about the use of DNA to unlock medical secrets and map human history. Science could now track each person's ancestry through the father's and mother's lineages, back generations or even thousands of years. Fascinating patterns of populations, their edges tattered by migrations and intermarriage, have survived over time, observable in our genome.

The TV program fired memories of the confusion Father Bill recalled feeling about his own ancestral identity. As a young boy, he had often sensed that he was somehow different. His family wouldn't eat pork, he and his sister spun tops on Christmas, and their parents had a strange ritual of sweeping dirt into piles in the middle of the room. They lit candles every Friday night. When a relative died and his family visited the home, all of the mirrors were covered. These were common practices for a few other Mexican Americans in Santa Fe, but not for most Catholics. Although confused and at times embarrassed, he never talked about this with his friends, even those from families that practiced similar rituals.

As he got older, he buried his curiosity. But the television program sent the memories flooding back. Where did he come from? He knew so little of his past, only that his family had migrated from Spain.

"I had always been interested in genealogy when I was a boy, but with all the Sánchezes in New Mexico, up until then I thought it would be pretty hopeless to track down my ancestors," Father Bill told a recent gathering at the New Mexico Genealogical Association. Emboldened by the ancestral treasury he's unearthed, he now shares the conundrum of the genetics of identity with all who will listen.

Spurred by the TV special, Father Bill located a testing firm in Houston, Family Tree DNA, which can track a person's paternal and maternal ancestry by identifying mutations imprinted in our genes. After receiving a kit in the mail, he and his dad swabbed the inside of their cheeks with Q-tips and sent them, slathered with their DNA, to the University of Arizona laboratory in Tucson that partners with Family Tree DNA.

"It was just a whim," he told the audience of about sixty people, packed with two dozen members of his extended family, some of them close relatives. "Frankly, we didn't expect to find much." After all, the Sánchez surname is far from obscure in the Spanish-speaking world. "In towns in Spain and Mexico, you can find Sánchez Pizza, Sánchez Sporting Goods, Sánchez Dry Cleaning, Sánchez, Sánchez, Sánchez all over the place, and none of them is related," he said to uproarious laughter by an audience of cousins and friends. There are thousands of Sánchezes in New Mexico, a state in which 40 percent of the 1.9 million residents trace their heritage to Spain. It's the equivalent of Smith or Jones to Anglos. And as with most Smiths or Joneses, those who share the name are not usually related.

A few weeks after sending in the swabs, Father Bill recalled for the audience receiving a call from Family Tree DNA that would change his life and challenge his identity. The precision of DNA to help distinguish blood relatives from those who merely share a common last name had everyone on the edge of their seats, listening to his surprising story.

"Can I speak to William Sánchez?" said Bennett Greenspan, the founder and president of Family Tree DNA.

"This is Bill Sánchez."

"We have the results of your DNA test," Greenspan said. "I wanted to call you personally."

With his company still small and genetic genealogy testing in its infancy, Greenspan would sometimes call clients when he found something unusual. He was surprised and excited when the results for William Sánchez crossed his desk.

"You're definitely of Semitic, of Middle Eastern, ancestry,"

Greenspan told Sánchez. "You have a marker that suggests that you carry a common Jewish mutation—a rare genetic marker that's found most commonly in Jews. If the Bible is accurate, you might very well be descended from Aaron, the Jewish priestly lineage that dates to the time of the Exodus."

Not many people are aware that Judaism has a modern priestly tradition. It dates to Exodus, when God instructs Moses to build the Tabernacle, a portable temple, in the desert, and selects the tribe of Levi—both Moses and Aaron are Levites—to serve in it. "You shall bring forward your brother Aaron, with his sons, from among the Israelites, to serve Me as priests," God says in Exodus 28:1. Aaron is anointed as the first high priest—the *Cohen Gadol* in Hebrew—in the line of Israelite priests, who were empowered to oversee the temple services. His male descendants are thereafter known as Cohanim, which like all tribal identity in the biblical era is passed along from father to son.

After the Roman general Titus sacked Jerusalem in 70 CE and put the torch to the Second Temple, the Cohen tradition withered in importance and became purely an oral one. According to Jewish lore, the Cohanim were directed to preserve their pedigree in anticipation of the return of the Messiah and the rebuilding of the Temple, after which they would resume their priestly functions. But one of the rituals of the Cohanim is still practiced today in Orthodox synagogues. During the recitation of the short prayer known as the Bircas Cohanim, which according to Numbers 6 the priests are required to say as a blessing of the Jewish people, a selected Cohen faces the congregation and forms a V sign with the fingers of both hands. The thumbs are extended and touching, which makes the gesture look like a W. That represents a *shin*, the first Hebrew letter in the word *shaddai*, which literally means "hovering over as a divine presence."

According to custom, the split fingers signify a channel for God to reach the congregation through the Cohen. The gesture should be familiar to hard-core Trekkies—devoted fans of the cult science fiction series *Star Trek*. Mr. Spock, played by Leonard Nimoy, would often give a "Vulcan greeting" to compatriots from his home planet. It consisted of the fingers of the hand split into a V. Invented on the spot by Nimoy in an early episode, it came to him, he says, from his childhood memories of the priestly blessing offered at the Orthodox Jewish services he attended with his family.

For those interested in biblical history, Aaron would turn out to be as giant a figure as Moses is to scholars of the Bible. Could it be true that Father Bill is a member of the tribe of Israel—more specifically, a descendant, perhaps apocryphal, of the line of Moses and Aaron? Although a last name like Cohen, Kohn, Cann, and the like may signal Aaronite ancestry, because of the relative newness of surnames, family names alone cannot confirm who is of Cohen lineage. About 3 percent of Jewish males today claim to be Cohanim. But until the development of genetic genealogy, there was no way to validate those oral claims.

That's now changed. DNA tests show that an extraordinarily high number of self-proclaimed Jewish priests, and some non-Jews with Middle Eastern roots, carry a distinctive marker on the male chromosome that dates approximately to the time Aaron was supposed to have lived. Like surnames, the Y chromosome is passed from dad to son 99.9 percent unchanged; it is not shuffled every generation like almost all our other genes. In principle, a son's male chromosome, with its distinctive markers, should be nearly identical with his father's, and father's father's, and so on, back to the lineage's original father—even back thousands of years.

This marker of Cohanim ancestry is a confirmation of Jewish fidelity and cohesiveness, if not definitively of true priestly lineage. For Father Bill, the news of his possible Israelite roots came as both a shock and a delight. He reveled in telling audiences the next part of his conversation with Greenspan. "He then said to me, 'What do you do for a living?'" The priest cracked a wide smile. The crowd, which had been tittering with fascination, burst into laughter.

"Well, I'll be," Greenspan had said when told of his profession.

For a moment he was unsure what to say, but then he quipped, "Well, we'll take you back!"

"Being a priest obviously runs in my family," Father Bill said as the crowd, Sánchezes and non-Sánchezes, roared.

THE CROSS IN THE STAR

Fascinated by his DNA readings, Father Sánchez launched into a study of the Hispanos (the term often used to designate Hispanics of Mexican American ancestry) of Santa Fe and Albuquerque for evidence of a Jewish past. Rumors have swirled for more than two decades that the American Southwest is home to hidden enclaves of crypto-Jews—Christians who practice Jewish-like rituals. Like many New Mexican Hispanos, his family practiced an amalgam of rituals, which he had long thought were drawn from only his Spanish and Native American ancestries. Many of his friends and neighbors, and thousands of Southwestern Hispano families, cover mirrors while grieving to prevent the spirit from escaping in the reflection, light candles on Friday night, refuse to work on the Jewish Sabbath on Saturday, avoid pork, and have newborn boys circumcised—all signature Jewish practices.

History now tells us that some Hispanos are descendants of *conversos*—the Jews who converted to Christianity during the late fourteenth and fifteenth centuries and then fled Spain and Portugal for the New World during and after the Spanish Inquisition, which began in 1492. But many converts secretly preserved Jewish rituals.

The Sánchez clan presents a provocative genealogical case study. More than likely, a group of *conversos*, who had previously converted from Judaism, left Spain around the time of the Inquisition, eventually settling in New Spain. Some maintained their Jewish customs or even practiced their faith in secret while outwardly professing Catholicism. When the Inquisition was imposed on Mexico in 1571, crypto-Jews found themselves fingered by their neighbors as apostates. Many fled for refuge into northern New Mexico and southern Colorado, where they eventually lost touch with all but the most superficial trappings of their Jewishness, embracing Catholicism or fundamentalist Protestant sects. Now, centuries later, their biblical ancestry is being revealed to them.

After getting his own results, Father Sánchez and his father, who subsequently died of cancer in 2003, launched the Nuevo Mexico project to study their extended family clan. He has since traced the male lineage of seventy-eight of his relatives and more than one hundred of his parishioners. Almost half of them have Semitic roots, with DNA markers that are very distinct from what is more commonly found in Spaniards. Of the seventy-eight relatives, thirty men, or 38 percent, have a Semitic marker only a few steps removed from those with confirmable Ashkenazi or Roman Jewish ancestry. Father Sánchez has genetically checked about a quarter of the men using a separate test that tracks their maternal ancestry. In almost every instance, the female lineage is Native American.

With rare exception, the Sánchezes who have genetic markers common among Jews revel in the findings of their probable Jewish roots. One of Bill's cousins, Norbert Sánchez, a deacon at a nearby church, Our Lady of Belen, recalls growing up in Jarales, New Mexico, south of Albuquerque, where local families would celebrate the "service of lights" each Friday night, often dining by candlelight.

"We always thought there was a Jewish background in our family, but we didn't know for sure. When I found out, it was like coming home for me," said Norbert. He proudly showed off a unique medallion hanging from his neck, a cross in a Star of David, which he calls the Sánchez coat of arms, during dinner at a local Mexican restaurant. "I'm Jewish by ancestry. But by faith, I'm a Christian, a Catholic. That's what I believe. I don't see any contradiction at all."

While most of the other local Sánchezes are content to just celebrate their enriched dual religious identity and remain firmly Christian by belief, a few have actually left their faith for Judaism. Maria Sánchez, a distant cousin of Father Bill's, was baptized and raised as a strict Catholic, never questioning her faith. But her family practiced many of the "Jewish" rituals common to other Southwestern Hispanos. They even played a game with a spinning top, similar to a dreidel. "My mother never talked about it as if this was anything other than normal," Maria explained. "We were too young to think it was strange. Since so many other families had the same rituals, we just assumed that's what all Catholics did."

As she grew older, Maria realized that she and a select few other local families were different. Her mother, Bersave, a devout Catholic, refused to talk about what she might know of her history, so Maria sought out her grandmother.

She remembers her looking frightened. "I asked her if we were Jewish. 'We're Israelites,' she told me. That's all she would say. She didn't want to talk about it either."

With the spell broken, her grandfather began to open up to her, reciting prayers that were an amalgam of Indian, Jewish, and Catholic traditions. "I learned the proverbs and prayers when I was knee-high. Grandfather knew," she said, "but he wouldn't acknowledge. It became an open family secret. I knew some of my friends who faced a similar situation, and we all handled it the same way. We knew but we wouldn't talk about it."

Maria was launched on a quiet, spiritual journey that has yet to run its course. In her forties, and after being trained as a home-based therapist, Maria is studying to become a pastoral counselor with hopes of becoming a rabbi. She has informally helped run services at Nahalat Shalom, where the local Reform rabbi welcomed Maria and other Hispano hybrids with open arms. The cantor, Lorenzo Dominguez, converted to Judaism years ago, when he finally came to terms with his Semitic ancestry.

Maria's mother, who still crosses herself, has had a hard time adjusting to the new mosaic of her ancestry. "There's a spiritual confusion," Maria said. "She's reluctant to turn away from what she's always believed in, and what was central to her was that she was Catholic." She occasionally goes to Jewish services and has adjusted some of her ritual celebrations, which always have had a Jewish undertone, to conform to the Jewish calendar. "And she was thrilled when I went to Israel, which really cemented my spiritual conversion," said Maria.

Any lingering doubts that Maria may have harbored about her newfound Jewishness were dispelled recently when her mother and sister both tested positive for one of the breast cancer mutations found almost exclusively in Jews.

There is a twist to this story, however: Maria has not yet formally converted to Judaism and has no plans to.

"It's not that I'm a believing Catholic," she said. "I just don't feel a need to convert, to fulfill the formal requirements. Why should I have to? I'm a Semite. I was born an Israelite. I see myself as a Hebrew. Judaism is in my blood, my genes. I've taken a journey to my Jewishness. Isn't that enough?"

For Maria and other Hispanos convinced of their Jewish roots, it's a familiar and plaintive cry. "I don't want to escape my Christian past. I want to incorporate it, to blend with it. I now proudly say, 'My family came from Spain and before that, we came from Israel.' That should be enough in the eyes of God, shouldn't it?"

Robert Martinez, one of Father Bill's cousins and an amateur genealogist, is baffled by Maria's spiritual conversion. He tested positive for the Cohanim marker, and there are also other, darker confirmations of his Jewish ancestry. At least eight people on his family tree have been struck by or died of rare genetic diseases found almost exclusively in Jews, and almost never in descendants of Spaniards or Native Americans. His sister Roberta, in her fifties, is the latest victim, diagnosed with breast cancer.

Has his genetic sleuthing influenced his religious convictions?

"Not at all," he said firmly. "I have Jewish roots, absolutely. But I know enough about DNA to know how problematic it is to choose certain slivers of your genetic blueprint and create a family myth out of it. My roots are in Spain; that was a country of countries, with people from all over the Mediterranean and Europe. America is a melting pot. I have French roots, Indian roots, Jewish genes, you name it. But I'm not Jewish. To me, Judaism is a religion, and I believe in Christ. I have Jewish ancestors somewhere in my past, I adore that, it has broadened my identity, but that's it. It doesn't influence my religious beliefs."

What does he think about Father Bill, whose discovery of his Jewish roots has shaken his commitment to Catholicism?

"It's good for Christians to recognize that Jesus was Jewish. It's a more personal connection to history. I gain something by having Jewish ancestors; I lose nothing. But I'm always a little nervous about people who suddenly want to become Jewish, or assume any identity, because of a DNA test."

Father Sánchez would later find out that although he almost certainly has deep Semitic ancestry, his Cohanim roots are in question. That's not shaken his quest to redefine his identity. When asked if he has ever contemplated formally embracing Judaism, there was a long pause that inched toward awkward.

"I don't need to." His voice, usually boisterous, turned into a whisper. "I never left Judaism. All of us, we're trying to unearth 'who we are.' We're going to have to define it for ourselves. That's a process that takes self-realization, dialogue, and self-reflection. For me, it's not just about what I believe. It's where I'm from. It's in my blood, in my genes. It's beyond choosing, by me. It's in God's hands.

"Knowledge of my Jewish ancestry has provoked me to question things, yes. Why did God choose the Israelites, the poorest of the poor, the most vulnerable? Why didn't he call the Greeks or the Romans, with their great art and cultural achievements? Why the Jews? I now question."

A quiet, dignified sense of betrayal enveloped him as he described the gauntlet Catholic Church officials have put him through in recent years as he's publicly shared his spiritual quandary.

"I have a pluralistic, not an antagonistic, view of our religions," he said. "I've been threatened with excommunication and suspension for talking about my Jewish roots, our Jewish roots here in New Mexico, but that does not deter me. I am drawing on my

Jewish consciousness," said Father Bill, who in 2005 fulfilled his dream of visiting Israel.

"I don't know where my Judaism ends and my Catholicism begins," he sighed. "I wish I, I wish those of us searching, had a spiritual leader who could lead us through this. If I had to choose between truth or Judaism or Christianity, what would I do? I'd choose the truth. After all, we are all the sons and daughters of Abraham."

CHAPTER 3

BLOOD TIES

Y ou alone I have singled out of all the families of the earth," God proclaims to Abraham, according to the shepherd Amos in Amos 3:2. His children are henceforth the chosen people. Why did God choose Abraham and the Israelites?

The Torah, the first five books of the Bible, never addresses that question. The reasons behind God's covenant remain beyond human understanding. All we know is that election does not turn on moral worth or rights but on duty, and it forever joined Jewish identity to blood ancestry, faith, and the gift of a homeland. The notion that the line of Abraham through his Israelite son Isaac is chosen has been a central tenet of Western religions since biblical times, even as the intellectual and political center of world Jewry shifted from Palestine to Babylon, Rome, Byzantium, Spain, the Ottoman Empire, Eastern Europe, and greater Germany, and in recent decades to Israel and North America. It forever ties Jews and Christians together. "As regards election, [Jews] are beloved, for the sake of their ancestors; for the gifts and the calling of God are irrevocable," wrote the Apostle Paul in Romans 11:28–29.

The claim of chosenness is vigorously contested by Muslims, who believe they have a unique connection to God through Abraham's Arab son, Ishmael, and his descendants. Both Jews and Arabs trace their literal lineages to pastoral nomads who later migrated to the ancient Levant (greater Israel) from Akkadia (Iraq), Arabia, the Egyptian delta, and the coasts of the Mediterranean Sea many centuries before the Hebrew Bible says God appeared to Abraham.

Abram, as he is first called in early Genesis, shows up in the Bible near the beginning of the second millennium BCE in the Mesopotamian valley. Today the rivers of the Tigris and Euphrates, the lifeblood of the region, run gently south past marsh grass, groves of palm, and fields of wheat and lentil before hitting the industrial center of Iraq, where they turn into smelly, shrunken brown sewers. It's believed that some ten to twelve thousand years ago, along these once-fertile shores, modern humans crossed a mysterious evolutionary threshold, as Paleolithic hunter-gatherers running in tribal packs became farmers. The emergence of agriculture and the domestication of animals led to a population explosion that would spread farming villages throughout the Middle East and eventually into Europe. The Garden of Eden, the cities built by Cain's sons, and the site of Noah's deluge were supposedly located in this region.

The Bible specifically refers to Abram's birthplace as "Ur of the Chaldeans," but there is no historical evidence of the Chaldeans for another thousand years or so after Abram's birth. Most scholars believe Ur recalls the great third dynasty of the Sumerians, founded by Ur-Nammu around 2100 BCE and located along the Euphrates River, not far from present-day Basra near the southern Iraqi-Kuwaiti border. Mesopotamia in the fifth or fourth millennium BCE was a magnet for nomads, drawing Sumerian-speaking peoples from the mountains or sea cultures and Semitic tribes that left the arid Arab Peninsula and migrated into the Mesopotamian greenbelt. Ur and the other thriving capitals in Sumer in the south and Akkad in the north gave birth to a rudimentary legal code, an irrigation system, and an elaborate trading network that extended to the mountains of Asia Minor and westward to the Mediterranean. Ships navigated the great rivers and traveled throughout the gulf, carrying pottery and various processed goods and bringing back fruits and raw materials.

The Sumerians, like other pre-Israelite communities, were poly-
theists who worshipped an array of humanlike gods. Archaeologists have found early examples of writing, which were apparently
the means for recording taxes and tributes to the ancient gods. Each
city-state housed a temple that was the seat of a major god in the
Sumerian pantheon, as thousands of gods controlled the powerful
forces that often dictated humanity's fate. Everyone had a duty to
please the local deity, not only for the goodwill of that god or goddess but out of respect for the other deities in the council of gods.
There are echoes of their religion in the Judeo-Christian-Islamic (or
Abrahamic) tradition, including the Genesis-like cosmogony of a
primeval sea that splits into heaven and earth, with the atmosphere
encased in a vault squeezed in between.

According to the Bible, about 2000 BCE, Abram's father, Terah,
becomes disenchanted with his life in Ur. The Bible does not offer a
reason why. Religious writers have speculated that he chafed under
the elaborate rituals and dictates of these pagan beliefs. Jewish legend holds that he owned an idol shop. One day, when his father
was on an errand, Abram smashed all the idols in the store except
for one, the largest. He put an axe in the surviving idol's hand and
awaited his father's return. Terah was predictably outraged and
even angrier at Abram's obvious lie that the axe-wielding statue
wreaked the havoc.

"You know those idols can't move!" Terah shouted.

"If they can't save themselves, then we are superior to them,"
Abram answered, "so why should we worship them?"

The Bible is less clear. For whatever reason, Terah uproots what
remains of his small family—Abram, his wife Sarai (later changed
to Sarah), and his grandson, Abram's nephew, Lot—and heads to
Canaan. He leaves a Sumerian civilization under attack by hostile
city-states and nature itself. Sumer had long been a granary for the
region, with bountiful fields of wheat. But as the climate turned
drier, farmers increased irrigation, which led to an accumulation
of salt in the soil. Although they gradually switched from growing
wheat to barley, which can better tolerate salt, within just a few

hundred years, the once-fertile plains of the Mesopotamian valley had begun to return to dust. Canaan beckons Terah and his family with its promise of an agricultural bounty on rich coastal plains.

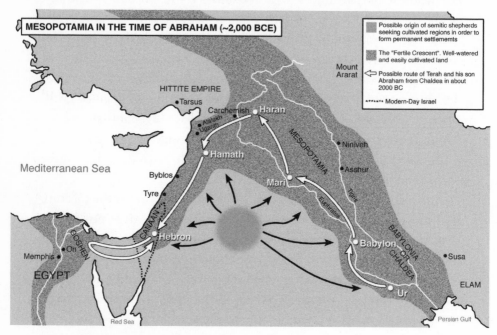

Figure 3.1. Abraham's trek to Canaan.

Terah does not make it that far. He passes through Babylon and Mari, two major population centers, before stopping in a caravan city on a tributary of the Upper Euphrates, where he later dies. Abram, a merchant by trade, remains there, living seventy-five unremarkable years, barren of children, before God first appears to him. "Go forth from your native land and from your father's house to the land that I will show you," the Lord commands him in Genesis 12:1–2. "I will make of you a great nation, and I will bless you; I will make your name great, and you shall be a blessing."

This entreaty prompts Abram's first act of unquestioning faith, a resumption of the family trek to Canaan. Abram and his family eventually reach the eastern shore of the Mediterranean Sea, where God reaffirms his miraculous promise to Abram and renames him

Abraham, the father-to-be of a "multitude of nations." It took many centuries before the ancestral line would crystallize into a people. "Abraham begot Isaac, who begot Jacob," and so it goes. The third in the line of Hebrew patriarchs, who is renamed Israel, fathers twelve sons, who become the forefathers of the clan of the Israelites, the original Twelve Tribes. Their descendants become known as B'nai Yisrael, the children of Israel—the Jewish people, bound together by faith, ancestry, and God's covenant, the gift of land. It would be a promise tested over many hundreds of years as the Israelites and their descendants wandered from their biblical homeland.

INTO THE DESERT

Prebiblical pagans were satisfied to find purpose in the sun, moon, the cycle of the seasons—the power of nature. Consider how appalled they must have been when they encountered these wandering Semites, who proposed that meaning could be found in something as insubstantial as One God. It was a crazy enough idea to change the course of history.

The story of Abraham is a central component of Western identity. How did the Jewish people, without a secure homeland for two thousand years, retain its cohesiveness, while so many other ethnic religions disappeared? These questions had sent me on my journey to where the Israelites first coalesced into a people—the Holy Land.

At sunrise, I headed by taxi from my hotel in Eilat in Israel to the Arava Desert crossing, north of where the city abuts Aqaba in Jordan. The two are joined in biblical history but forever divided by geography, religion, and politics. The border separating the countries is demarcated by a series of low mountains and dusty valleys formed during the Pleistocene epoch, some fifty million years ago, when the earth's crust buckled and boiled, spewing forth lava and rocks. Tectonic plates fractured and smashed up against one another, ripping the megacontinent into what is now Africa and Asia. With the freezing and thawing of countless ice ages, the earth sliced a 4,000-mile-long gash, 20 to 60 miles across, known as the Great Rift Valley. This geologic depression stretches from the mouth of the Zambezi River in Mozambique northward through East Africa, the Jordan Valley, and into Syria. Where the continents separated, chasms like the Dead Sea Valley resulted; where the continents collided, mountains formed, such as Mount Sinai in Egypt, Israel's Mount Hermon, and those surrounding Petra in Jordan. Bizarre rock formations and an abundance of underground springs mark this terrestrial rip.

It has long served as a crossroads—for the first modern humans who passed through on their way out of Africa tens of thousands of years ago and for the periodic march of warriors who have left evidence of their conquests forever since. Over hundreds of thousands of years, innumerable peoples have staked competing claims to this sacred part of the world: Egyptians, Phoenicians, the mysterious Sea Peoples, Philistines, Assyrians, Babylonians, Persians, Greeks, and Romans. But only the Semites sank their roots deep in the treacherous sands and have endured. The Bible says Moses accepted the inscribed tablets that would provide the moral foundation of Western civilization on its frontiers. This is where Jesus was born, ministered, and martyred. Muhammad ibn Muhammad, the prophet of the Muslims, passed through here on his pilgrimage from Mecca to Syria.

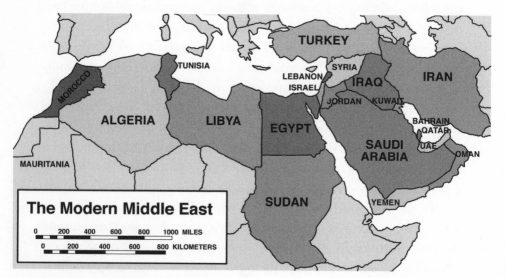

Figure 3.2. Israel in the Arab world.

In the afterglow of the Oslo Accords, signed in 1993, the border had bustled with commerce and tourism. On this early morning, when an Israeli taxi dropped me off at the border gate, I was the only visitor. A half-dozen Israeli guards and inspectors, rifles slung lazily over their shoulders, eyed me warily as I pushed my luggage cart across the paved road that spans the border, widened just a few years ago to accommodate the anticipated crush of tourists. The Jordanian side, recently spiffed up, was deserted, except for a rumpled figure watching me approach.

Walid Al-Mallah, my guide whom I had hired through a Jordanian travel agency, greeted me with a polite but reserved smile. He is a small, thin man, in his midforties, with a droopy mustache and sad eyes. Dressed like a college professor in a tattered tweed jacket, frayed shoes, and a floppy safari hat that barely covered his thinning hair, he looked much older. We hopped into his car, a Toyota that was battered inside and out.

Aqaba, like Eilat, is a key industrial port as well as a resort destination. Walid snaked along the gulf past the heavy trucks and tankers ferrying phosphates and other minerals for export before turning north and east. Within minutes, the sea vistas gave way to the brown and gray hills that loom over the harbor, which then yield to the desert. Gradually, over the course of a two-hour ride, the rocky terrain tumbles into the windswept sands of the desert on the outskirts of the biblical Middle East.

The vast, sandy plain known as the Wadi Rum, the massive dry riverbed that cuts across southern Jordan, was a major byway in biblical times. "Only God could have created this," Walid commented as we drove past the extraordinary series of weird ridges and bulbous domes that look like they were made from candle wax melted in the burning sun, almost like buildings designed by Antonio Gaudí. Formed by the sequence of faults that carved the Rift Valley, the Wadi Rum has been described as lunar, but given its prevailing reddish color, "Martian" is more apt. Enormous bronze-hued sandstone formations jut into the sky, a symbol of nature's power that leaves one breathless. "Vast, echoing, and God-like," the English adventurer T. E. Lawrence rhapsodized in his book *The Seven Pillars of Wisdom*. In the distance, thrusting skyward from the desert floor, rising 2,500 feet or more, are the seven legendary columns that welcomed Lawrence and the Arab army during the rebellion against the Ottoman Empire during World War I, one of an endless number of battles that have shaped the region's fate.

We pulled into Rum village for lunch at the café at Abu Aineh, from where the valley unfolds. We finally broached the topic on everyone's mind, every day, in Israel and Jordan: Is there any hope for peace between the Jews and Arabs? The question can be framed in another way: Is there common ground on the issue of the Right of Return to biblical Israel—the belief by Jews and Palestinians that they each hold spiritual and political claim to this disputed territory? The contested covenant illustrates the entangled narratives of religion, politics, and science that haunt the Middle East.

"Look around," Walid said, pointing to the sandy square across from the café. It was deserted except for one lone couple that stood posed and smiling in front of a bored camel that appeared to spit on them as a Bedouin snapped their picture. "This place should be hopping. Instead, all we have is fear. Before the violence, it was packed with tourists lining up for camel and jeep rides through the Wadi. We're nowhere near any danger areas, but people are scared, too scared to visit."

Although Walid is a Muslim, like many Jordanians he lives a secular life. He holds a degree in archaeology from Jordan University. But years ago, with jobs scarce, he turned to giving tours to support his wife and two children in Amman. It was lucrative work for a brief few years in the 1990s, after the borders burst open, but now he scratches to survive. It was November, and I was only his fourth tour of the year. As Walid and I talked, we downed a ritual cup of a bitter greenish liquid euphemistically known as coffee that's popular with Bedouins. It's an acquired taste. Curiously, the high-voltage caffeine jolt dissipated the lingering nervousness between us.

Walid finds the fundamentalism of radical Islamists and the intolerance of extremist Jews equally distasteful. "I've met too many different kinds of people here—Arabs, Israelis, Christians—to believe that any spiritual path is superior," he said. "We all could share in this land. This valley, Jordan, Israel . . . this is a special place. Arabs and Jews know this. Yet, we haggle, over what? Land. The terrorism, the killing . . . it's a fight between those who passionately hold differing views about their heritage, yet we are of the same blood. These are two stubborn peoples."

Later that afternoon, after a mesmerizing camel walk and jeep ride across the wadi, Walid led me to a tranquil Bedouin encampment at Ain Shalaaleh, also known as Lawrence's Spring. It is on the *deereh*, the vast tribal territory that lies on the routes used by Islamic pilgrims trekking by camel and foot to Mecca until the early twentieth century, when trains replaced camel caravans as the principal mode of travel. The desert oasis is home to a dozen families that caravan through the desert, much as their ancestors did for thousands of years. The Bedouin are camel breeders, which once offered a degree of self-sufficiency in the harsh desert. The camels provided milk and clothing made from their hair, and their urine was used as hair-washing lotion. When times were tough, they could be slaughtered for meat or skin used to make leather for the round huts that remain a Bedouin trademark. Some 300,000 of Jordan's 4 million Arabs still live as nomads, many in portable camps like this one.

Walid had arranged for *shy*, a kind of tea, with Abou Mohammad, the leader of the encampment. The Bedouin are renowned for their graciousness. Abou warmly grasped my hand and guided me past his favorite camel and a penned flock of goats and into one of the sturdy tents of his transient village. Worn but colorfully braided cushions covered the dirt floor. Two little boys, giggling and prancing around in shorts and bare feet, scooted out of the room as we sat down cross-legged. He shouted something in Arabic. An instant later, his wife appeared from behind a curtain, carrying a teakettle with shot glasses, and then quickly left without making eye contact.

Abou poured the tea, scalding hot and dark yellow, into our glasses and proposed a toast. "To our past and to our future," he said with a wide, broken smile, his teeth browned by strong teas and nonexistent dental care.

I curled my lips as I sipped, the bitter tannins eventually yielding to a sweet infusion of sugar; the second sip tasted great.

When I told Abou I had come to explore the history of the Holy Land, he became very animated. His eyes, set deep in his leathery face, dark but not hard, shone in the shadows. Walid acted as the translator: "The Bedouin are proud of their heritage. There are thousands of our tribe in Jordan, Saudi Arabia, and Kuwait." Walid took pleasure in embellishing Abou's story. "They believe they are the children of Shem, son of Noah," he said, citing the story in Genesis. "That makes them Semites, as are all Jews and Arabs."

Then Abou interjected in almost perfect English. "There are studies. There are studies." For a moment, I was taken aback. Had I heard right? Could he be referring to the subject of my visit to the Middle East? Had word of the research into the genetic origins of Arabs and Jews, which made headlines in London and New York, somehow crossed ocean and desert to the Wadi Rum?

Walid confirmed my speculation. Yes, Abou had heard of the DNA research. As it turns out, Abou has relatives in England and the United States. The story had also been widely reported in the Arab and Israeli media and on BBC Worldwide. He proudly took out his shortwave radio and spun the dial, the radio spilling out a cacophony of languages. Why should I have been surprised? After all, the Bedouin working at the tourist center had cell phones clipped to their waists. And there were satellite dishes in the cement-block village a few thousand meters away, where some of Abou's tribe lived in crumbling concrete houses.

Talk of a global village no longer seemed like a cliché. Here we were in God's country, literally, and the discussion had turned to the Bible and genetics. Abou recalled that in the halcyon days before violence between Israelis and Palestinians erupted in 2000, he often had discussed biblical history with visiting Israelis. He said

In genetic terms, the Hebrew Bible is the story of the Y chromosome. Although Jewish tradition has traditionally emphasized the importance of the female lineage—according to Orthodox belief and the laws of the State of Israel, to be a Jew, one must be either a proselyte or the child of a Jewish mother—the male lineage was the tribal standard of the ancient Middle East.

"This is the record of Adam's line," reads Genesis 5, which tells the story of the creation of humankind. Chapters 10 and 11 relate in rote fashion the male descendants of Noah, and the world's "races" that grew from his seed. With its exhaustive ancestral lists, the Bible qualifies as the world's first ancestral primer. "All Israel was registered by genealogies," declares 1 Chronicles, which goes on to list all the male descendants of the line of Adam over thousands of years—"the chiefs in their families"—which extends to Abraham, the first Hebrew and the father of the Israelites, and forward to Moses and Aaron.

Could anyone really hope to trace their ancestry back dozens of generations to biblical times? Could a distinct lineage have been maintained throughout the long exile of the Jewish people? What would confirming this connection mean to Jewish identity?

These questions turned in my mind as I jumped on a train in Tel Aviv for Haifa to visit Skorecki. His office sits along the Mediterranean shore at the Israeli Institute for Technology. Technion, as it is called, is anchored by the Rambam Hospital, named after Moses Maimonides, the great Jewish physician of late-twelfth-century Córdoba, in Spain, who drew his inspiration from Greek and Roman thinkers, wrote in Arabic, and embraced the wisdom of the three great Western religions. Maimonides was known by the acronym of *Ra*bbi Moses *b*en Maimon—Rambam.

Like the Spanish rationalist, Skorecki, who is in his midfifties, is an unusual—some might say heretical—mixture of hard-nosed scholar and Orthodox believer. He greeted me with a warm handshake, ushering me into his window-walled office high atop the faculty building. Although unassuming in appearance, with carefully trimmed short hair, large round glasses, and a boy-next-door niceness about him, there is nothing understated about his impact on genetic genealogy. He has dual appointments as head of nephrology and molecular medicine at the medical school and the hospital, but he spends an increasing amount of time overseeing DNA research projects.

An intensely driven man, Skorecki is motivated as much by his commitment to Israel as by scientific curiosity. He was born in Canada and became an Israeli citizen more than a decade ago. He is what is called in Hebrew an *oleh*, which means "one who goes to live in Israel." But the word suggests much more. It's from *aliyah*, which literally means "to go up to" and implies the "moral and spiritual superiority of living in Israel." Aliyah is enshrined in Israel's Law of Return, which grants any Jew the right to assisted immigration to Israel, as well as automatic Israeli citizenship.

Skorecki's office commands a magnificent view of the Mediterranean coast. "That area is controlled by the Hezbollah," he said, pointing northward to southern Lebanon. The Hezbollah are a militia group of radical Lebanese Shiites, which formed after Israel occupied Beirut and southern Lebanon in 1982. Their suicide bombings against the Americans and French in Beirut in 1983 initiated the demonic era of suicide bombings—or "martyrdom operations" as its proponents call them. They later aligned themselves with Palestinian nationalists and have been an implacable enemy of Israel. Hezbollah took control of the border region when Israel pulled back its troops in the spring of 2000, and the Lebanese government refused to replace them with its army. Hezbollah's border incursion and renewed bombing of northern Israel in the summer of 2006 ignited a major confrontation with the Israelis that left thousands dead.

"They occasionally shoot their Katyusha rockets into our encampments along the border, sometimes even into our towns," he said, almost as an aside. Impeccably polite almost to the point of shyness, Skorecki is not one to make his points by overstatement. Like many Jews who live with the daily threat of a Hezbollah missile strike, he is alternately stoic, anxious, and combative, but without ever losing sight of his commitment as a doctor to his people. "This is a vulnerable area. The missiles are a constant worry. Part of my brain is always thinking about my family. But I have to focus on our mission, our responsibilities as Jews."

Skorecki, the only child of Holocaust survivors, defines himself by his Jewish identity. "It's metaphysical and not physical," he said. "Our identity is based on an oral tradition, law, culture, custom, and not on physical attributes, including DNA. But genetics can tell us a great deal about origins and common ancestry. It can help us piece together how we survived over the centuries as a people."

Before World War II, Skorecki's father, Eli, one of nine children, lived in Russian-administered Poland before moving to Kraków. Eli was poor, began work at age eleven, and never got much of an education beyond some Jewish instruction. When the Nazis came, he was shuttled to the ghetto portrayed in the book and movie *Schindler's List*, then to a concentration camp in Germany, before American troops liberated him.

Skorecki's mother, Sabina, was more educated, having been raised with her parents and four brothers and sisters in a village in Poland that until World War I was part of the Austro-Hungarian Empire. But the full weight of Hitler's Anschluss (the annexation of Austria by Germany) in 1938 descended on her family. They were swept out of their homes and into the infamously harsh concentration camp at Mauthausen. Everyone perished but Sabina. She ended up in Auschwitz, where she survived for twenty months before being freed by Russian forces.

Eli and Sabina met in Kraków after the war and soon married. A few years later, they moved to Canada to escape the devastation and brutal memories that haunted them—and would so profoundly influence their young son growing up in middle-class Toronto. The Holocaust, which shattered so many people's faith, deepened Karl's. Seared into his memory are his mother's stories of the terror of the selection process by which some Jews were sent to the gas chamber while others were given a temporary pass to another day. "She remembers being in a bunk with women, and a number of them knew they were going to be killed," he said. "I always thought that when you were selected, you were sent right to the gas chambers. I didn't realize the numbers were written down and for a day or two they knew they had been selected for extermination and had to live with that." His parents' experiences set him apart from his fellow Jewish students, who could not comprehend his family's devastation. "I became more observant through the years, mostly because of my parents' stories and the influence of my teachers, hearing about Jewish persecution that was close to me."

By the time he turned eleven years old, young Karl was spending his summers at Zionist camps in Israel and learning to speak Hebrew. His other passion was science. "In grade six or seven, one of the first essays I had to write was about evolution," he recalls. "It was in Hebrew. I remember the rabbi was very disturbed when he read the title. He asked me to change it, and I didn't know why. Only years later did I realize that the Hebrew word for 'sex' and 'species' is the same. My point is that there is no great incongruity between science and religion. It's all a matter of understanding, of context. The first chief rabbi of British-occupied Palestine wrote that the Bible is consistent with evolution. I think even modern Orthodox Jews believe that, except for what I call 'literal fundamentalists,' who are rare."

Skorecki has always been especially proud of his Cohen lineage. At his bar mitzvah at age thirteen, he was excited to be called to the Torah to offer the priestly blessings reserved only for Cohanim. "It was an honor," he said. "My father would tell me about his father and his brothers participating in the special traditions. I had anticipated for years becoming a part of that." Without the Temple standing in Jerusalem, the privileges of today's Cohanim are limited. In practice, being a Cohen is more an affirmation of a tradition than a responsibility. Leading the congregation in prayer was a special moment for a devout yeshiva student whose Jewish identity had been shaped by his family's travails and sealed by visits to his religious homeland.

At fifteen, Skorecki decided to pursue his studies in medicine and pledged that he would eventually move to Israel. Even while training as a nephrologist at the University of Toronto, he continued to follow Aaronite traditions, including the restriction on touching dead bodies. "It required some negotiations with my professors," he recalled, chuckling. He spoke to a rabbinical authority, which gave him special instructions on a variety of issues, including how to handle corpses with rubber gloves.

Skorecki's dream to make aliyah was delayed for more than a decade by his responsibilities as a son to his aging parents and as a father to a young family. He established his practice in Boston and then moved back to Canada, taking a position as a professor at the University of Toronto. In the early 1990s, during a sabbatical in Israel, he and his wife, Linda, finally made a firm commitment that they would raise their children in their biblical homeland.

Just months before leaving North America for good, he found himself sitting in a pew in a Toronto synagogue, contemplating what it meant to be a Jew. He does not embrace the literal truth of scripture, but he is spiritually inspired by it. Why not see if he could track Aaron's lineage? As a scientist, Skorecki was aware of the exciting application of DNA analysis to disease and population research. Geneticists had already published a number of comprehensive studies using classical genetic markers, such as blood proteins, suggesting that the diaspora communities of Jews had deeper common roots than anyone could have imagined. But his interest was far more targeted: he wanted to know about one specific genetic lineage—his own paternal ancestry, supposedly shared by an exclusive fraternity of Cohanim.

As the Sephardic Cohen began reading his Torah portion, Skorecki speculated that if all Cohanim, including himself, were indeed descendants of one man, they should have a common set of genetic markers, which originated with their shared distant paternal relative. If Jewish tradition had held firm and religious Jews were faithful, the priestly inheritance—the lineage of the Levites and especially the subgroup of high priests, the Cohanim—should be relatively intact because almost no recombination occurs on the

Y. If his hunch was correct, all Cohanim should share a marker for a common ancestor who lived in the Levant during biblical times.

"I like clean sorts of questions that you can get closure on with a study," Skorecki told me. "I thought this would be a great summer project for a medical student. That was the idea." He underestimated its allure. This elegantly simple inspiration would soon ignite the imagination of the world. It could be evidence of Jewish lineage that traced back to the Exodus.

THE FIRST ISRAELITE PRIEST

The Hebrew Bible is the only record we have of Moses and Aaron. If they lived, they were probably born in Egypt between the eighteenth and the fourteenth centuries BCE. Highly literate, the Hebrews thrived in their new home. "The Israelites were fertile and prolific; they multiplied and increased very greatly, so that the land was filled with them," reads Exodus 1:7. Scholars speculate that the once-bickering tribes morphed into a cohesive people by making themselves indispensable to the Egyptian monarchy, providing services ranging from bookkeeping to resolving legal disputes and establishing themselves as essential intermediaries between the powerful and the powerless—a role their descendants, the Jews, would assume throughout their history. According to the Bible, this golden age in exile ended abruptly when a new pharaoh revoked their privileges and forced them into bondage.

Although his historicity may be in question, Moses plays a seminal role in Western religion. He is the man chosen by God to free the Hebrews and to receive His laws. "I have come down to rescue them from the Egyptians and to bring them out of that land to a good and spacious land, a land flowing with milk and honey," God informs Moses in Exodus 3:8. Moses does his best to convince the Lord that he's not the right man for the challenges that lay ahead, but God prevails and includes Moses's older brother, Aaron, in the plan. After a series of miracles performed by Moses fails to convince the pharaoh to release the Israelite slaves, the Lord unleashes a torrent of plagues, which is supposed to culminate in the Passover slaughter of every firstborn in the land. With the children of the pharaoh facing death as well, the pharaoh relents and agrees to let the Hebrews go. Moses and Aaron gather the flock and set off across the Sea of Reeds to begin their epic sojourn. The pharaoh then abruptly changes his mind and gathers the chariots of Egypt to pursue them. All that night, a strong east wind blows and raises walls of water that protect the Hebrews but ultimately crash down upon the Egyptians.

Barely three months into their wanderings, God summons Moses to Mount Sinai, where he announces in Exodus 19:5–6 a codicil to His covenant: "[I]f you will obey me faithfully and keep My covenant, you shall be My treasured possession among all the peoples . . . a kingdom of priests and a holy nation." Election is contingent on the Israelites fulfilling God's ethical and religious prescriptions, which subsequent rabbis and scholars would later codify in Jewish law, the Halakhah. God designates Aaron as the high priest of the Israelites, to which the DNA findings may bear witness.

Following the Exodus, the rest of the Levites, including Moses's descendants, were assigned the special duties of carrying the Ark of the Covenant and serving in the central sanctuary under the direction of Aaron's descendants, the Cohanim. For the first thousand or so years of Israelite history, Jewish priests had an important religious role. Only Cohanim were allowed to make sacrifices, burn incense, and offer blessings and other rituals. Only Cohanim were allowed inside the Tent of Meeting, or Temple. "Any outsider who encroaches shall be put to death," reads Numbers 3:10. Only the Cohen Gadol himself could enter the Holy of Holies, the Inner Sanctum, on Yom Kippur, the most sacred day on the Hebrew calendar.

The Aaronites lost their priestly status after the first century of the Davidic Empire, only to reclaim it after the return of the Israelites from the Babylonian Exile and the rebuilding of the Temple.

In announcing the new order, the biblical prophet Ezra, a Cohen Gadol and scribe who was a member of the Great Assembly, the ruling body for more than a century, decreed that only *Cohen M'yuchas*, Cohanim of verifiable lineage, would be allowed to perform the priestly functions.

From the time of Ezra onward, Cohanim became subject to special laws to preserve their status. They could not marry a convert, a "harlot," or a divorcée. They could not touch a corpse or even enter a cemetery. In later centuries, only Cohanim thoroughly investigated by the Sanhedrin, the Jewish high court in Palestine during Roman times, were allowed to participate in the Temple service. In exchange for guarantees of spiritual purity, the Cohanim and in particular the Cohen Gadol were granted special privileges, including a *trumah*—a share of the produce raised—and some of the offerings from the Temple sacrifices. Numbers 18 calls the high priesthood "a service of dedication," which comes with sacred perquisites "for all time"—the best meats (except for firstborn cattle, sheep, or goats, which are consecrated) and "the best of the new oil, wine, and grain." The rest of the Levites, the junior priests, were assigned a lesser share of the tithes, one-tenth of which was to be set aside for the Lord.

As compared to the great patriarchs and prophets, Aaron is a minor biblical character, and his history as the father of the Jewish priesthood is a footnote in Jewish history. Many Jews do not even know of the tradition. But for devout Jews, Muslims, and Christians, Aaron's story is more than just a good yarn; it is an accounting of a brother's faith and a mark of the sacred devotion of an entire family to God's calling. But is there evidence of it in our DNA?

THE Y TEAM

Within days of his inspiration, Karl Skorecki dove into the articles on genetic genealogy to see if his idea was even feasible. "There was all this literature at that time saying the Y chromosome was sort of hopeless," he said. But the science was evolving quickly. Unsure exactly how to proceed, he contacted a population geneticist in Israel whom he had befriended on one of his frequent visits. "I faxed a note to Batsheva Bonné-Tamir in Tel Aviv to get her thoughts," he said.

Bonné-Tamir is an emeritus professor of genetics at Tel Aviv University and the former director of the National Laboratory for the Genetics of Israeli Populations. A shy woman but a bold thinker, she is by general acclaim one of the matriarchs of Jewish genetic research. Would it even be possible, he wrote the legendary geneticist, to find genetic markers to distinguish between Cohanim and the descendants of the common Israelites? Were there enough identifiable mutations on the Y to yield such information?

"If she didn't think a Cohanim study could work or had serious reservations about it, I wouldn't have gone forward," Skorecki said.

The field of genetic anthropology did not even exist in 1932, when Batsheva Bonné-Tamir was born. Her parents had come to Palestine eight years before amid the initial flood of immigrants heeding the Zionist call to repopulate biblical Israel. "Since I was a child, I've been interested in how nature and society interact," she told me. She noticed the dramatic influence of genes even among people raised in nearly identical environments. "I was always curious how siblings could be so different." Her sister Eve is short; she is much taller. Her sister is more controlled and went on to become a psychologist; she is more emotional. "My mother used to say she wanted to pin us together to make a whole person. It was one more reason why I wanted so much to study biology and the environment."

After a time in a kibbutz and a stint in the Israeli army, Bonné-Tamir began her studies at Hebrew University in Jerusalem. She remembers being struck by the vast differences in the stature, hair and eye color, and even demeanor of the new immigrants. They looked and acted so differently, but they all claimed to be descendants of the ancient Hebrews. "Israel is a genetic fishbowl with its colorful mixture of dozens of populations," she said. "I wanted to study nature and nurture, but there was a great divide between the sciences and social studies at the university at the time. You were forced to choose. I was undecided, but eventually chose the social track. I was lucky enough to study under Elisabeth Goldschmidt."

Professor Goldschmidt and her friend and colleague, Chaim Sheba, who separately fled Hitler's Germany, were true pioneers of Israeli science. They trained the first homegrown generation of population geneticists. Bonné-Tamir was their prize student. Since the 1960s, when she wrote a brilliant doctoral dissertation on the genetics of the Samaritans, a tiny religious and cultural minority in Israel and the West Bank that claims to be descendants of the ancient Israelites, her studies of the ancestry of Jewish and Arab ethnic populations have literally defined the field and helped transform a country no larger than the state of New Jersey into a major center for genetic research.

The message from Skorecki sent Bonné-Tamir's imagination soaring. She recalls being almost unable to contain her excitement. "I was immediately taken by Karl's idea," she told me, as we sat chatting in her office at the Sackler School of Medicine. "It was a simple and an ingenious idea for a study."

She unearthed the fax reply that she had sent back to Skorecki. Bespeaking her unfailing courteousness, her handwritten note included an apology for its not being typed—"I didn't want to wait for secretarial help till next week," she had jotted excitedly in the margin. "I find your ideas regarding application of genetics to Jewish Genealogy quite attractive." Here was an opportunity to examine the fate of one of Western civilization's oldest lineages! Here was a chance to test the belief that the Jews of modern Israel were actual descendants of the ancient Hebrews, returning to reclaim their homeland after centuries in the diaspora!

Bonné-Tamir reassured Skorecki that the Y chromosome might not be the barren landscape that most of the scientific literature made it out to be. To address that challenge, she said, "You need to talk to Michael Hammer at the University of Arizona."

It was February 1995, just a few weeks after Skorecki's brainstorm. "I remember getting an e-mail from Karl and then a telephone call before he moved to Haifa," Hammer recalled. "He laid out the idea. He said, 'I'd like to test whether we can find markers for Aaron's descendants, and you're the expert on the Y. What do you recommend? Can we do this?'"

Hammer is as Reform and secular in his Judaism as Skorecki is religious; as casual as Skorecki is formal. What links them is their curiosity about the origins of Jewish identity. "To have Karl raise this issue about the Cohanim, as a Jew, it intrigued me," he said. "As a scientist, I could ask on a microevolutionary scale if in fact the Jewish priesthood was really passed down from father to son with great care. I didn't have a good sense if I could ever think of Judaism in terms of roots as maybe an African American would think of roots and what part of Africa they were from. I think I've always had this sort of question ongoing in my mind about my Jewishness, the issue of 'What is a Jew?' So, this was an opportunity for me. It was something we as scientists rarely have the opportunity to do, sort of like 'evolution in a test tube.' Karl's idea got me very interested in what the Y chromosome could tell us."

"When I contacted Michael, all I wanted at first was information about how to amplify the genetic markers on the Y, and he was kind enough to supply it," Skorecki recalled. "But he told me there was another person, in London, who was interested in Y chromosomes and Jewish populations, and this person had contacted him." About the same time as a kidney specialist in Toronto was musing about Aaron's genetic legacy, a worm aficionado turned accountant cum geneticist in London had been asking himself much the same question.

"Why don't you call Neil Bradman?" Hammer suggested to Skorecki.

* * *

Neil Bradman is a stocky, jocular man with a high energy level, a
sharp tongue, and an even sharper mind. When Skorecki rang him
up, Bradman was still a doctoral student, although an unusual one.
He was a businessman in his fifties, who had studied worms years
ago while in graduate school, before abandoning his academic interest in science to become an accountant. He became a millionaire many times over and still runs the British publishing company
and small real estate empire that he cofounded decades ago. But in
the early 1990s, his interests took an abrupt detour. Within a few
years, Bradman would transform himself into one of the most influential genetic archivists in the world and eventually take over as
chair of the Center for Genetic Anthropology at University College
London. His very first major project remains his most famous: the
study of the genetics of the Jewish priesthood.

"I had no idea about Karl's idea or the research project he wanted
to launch," Bradman told me. "It was quite an accident, really, that
I tripped onto the same question. Studying Jewish priests was a
summer project idea for my son. Up until then, I would have been
happy enough studying nematodes."

Bradman's obsession with parasitic worms dates to his days at
London and Liverpool Universities in the 1960s, when he studied
zoology and parasitology. But he abandoned his science research
for the more financially promising world of business. "I always
wanted to work on genetics, but after all, I enjoy eating," he said
with his characteristic booming laugh.

Twenty-five years later, he found his past tugging at him. "I may
not have started out as a geneticist, but I breed geneticists," he said
with a chuckle. His two oldest children, Avi and Robert, whom
he affectionately calls "son number one" and "son number two,"
were both studying genetics at Leeds University. Their research re-
awakened his dormant interest in science. "I reread the textbooks
and got a bit of a vicarious genetics degree."

Wealthy and secure, he decided to follow his youthful passion.
His first thought was to shadow a doctoral student and renew the
study of his beloved worms. But he abruptly changed course. He

credits romance and religion. Shortly after he began refocusing on science, in the fall of 1994, his son Robert came to him for advice. He was searching for a third-year project to complete his graduate genetics degree. The only requirement, Robert said, was that the research had to be in Israel, because he was then dating an Israeli (whom he would later marry).

"We sat down, and it all came together in a flash," Bradman recalls. "In about five minutes we had come up with this idea of testing the story of the Jewish priests." He rang up the one person in Israel who he knew would level with him about whether the idea was even workable: Batsheva Bonné-Tamir. It was the second call she had received on this offbeat idea in a matter of weeks, and she offered the same advice each time: contact Michael Hammer in Tucson.

Bradman was soon at work collecting samples when he received a surprise call from Skorecki, who had serendipitously come up with the idea on his own. The circle was coming together.

Although Bradman does not share Skorecki's fervent religious beliefs, he is a practicing Jew and a Levite by oral tradition. "I'm a park-the-car-around-the-corner attendee of an Orthodox *shul*," he joked, referring to the fact that religious Jews are not supposed to drive or do anything resembling work on the Saturday Sabbath. But there is no question that his interest in genetics is a consequence of his Jewishness. "I want to use it to uncover the history of communities and to give voice to the silent mass of people who may not have left written ethnographic or archaeological records," he said. "And many of those lost communities are Jews."

THE FIRST COHANIM STUDY

Bradman flew to Israel and met with Skorecki at the Nof Hotel in Haifa. They traded ideas about how to proceed and arranged for Bradman's son, who often traveled back and forth to London, to help keep them both up-to-date. They needed more samples from

Jews of diverse ancestry—and there was no better place to find
them than in Israel. With the steady tide of immigrants who have
washed onto its shores over the past century, Israel offers one of the
richest sources of ethnically and geographically eclectic DNA in the
world, concentrated in a compact area.

Stocked with what looked like tubes of mouthwash, Robert
Bradman set up a table at the Western Wall in Jerusalem during
the Jewish High Holy Days in the fall of 1995, where he was sure
to find an ample supply of Jewish priests. At the Jewish New Year,
Rosh Hashanah, and the solemn Yom Kippur celebration that occurs ten days later, thousands of Jews gather to pray aloud before the sacred remnant of what was once the Temple of the Jews.
Closest to the wall, rows of Jewish men of all ethnic backgrounds
lift their prayer shawls over their heads to form billowy canopies.
The benediction Bircas Cohanim, recited at every Jewish holiday,
rings out across the square. "May the Lord bless you and keep
you. May the Lord make His face to shine upon you" can be heard
in any number of languages. On those early fall days, after the
prayers, the researchers gathered off to the side as many as would
participate. The Cohanim swished and spit, filling tubes with DNA-
bearing saliva in an exercise that looked like an orchestrated
communal dental cleaning.

The newly assembled research team examined the Y chromosomes of nearly two hundred Jewish males—half of them Ashkenazim and half Sephardim, a third of whom claimed to be Cohanim.
If the Jewish tradition held true and modern Cohanim did share an
ancient ancestor, there should be evidence for it in the form of an
ancestral haplotype—a set of DNA markers. Over the years, as
the genetic family tree sprouted more and more branches, more
microsatellite mutations would have been imprinted on the signature haplotype, but one dating to the time of the Exodus should
still be identifiable.

What did they find?

Embedded in the data was pure dynamite: almost every one
of the Cohanim, regardless of whether he came from the Middle

East, India, Africa, Europe, or the Americas—98.5 percent of those tested—had a signature mutation pattern. The marker was found in only about 3 percent of the general Jewish population. Hammer was stunned. "At first I was worried that it was a sampling artifact, a chance result of the group chosen, because of the high frequency," he said. He was even more startled to find one haplotype common to both Ashkenazi and Sephardic Cohanim, indicating that it pre-dated the split of Jews more than a millennium ago.

When a letter summarizing the research was published in *Nature* in January of 1997, the researchers felt confident enough to float a bold inference. "This may have been the founding modal haplotype of the Jewish priesthood," they wrote.

While they were all personally fascinated by the findings, no one anticipated that the study would turn them into international celebrities. "At first, I didn't think it would be of interest to anyone but scientists," Hammer says.

Skorecki had been even more dubious. "I really did not even think that this would ever get published in *Nature*. I thought this was much too parochial a story. I'm not all that genealogically ori-ented," he added. "I have the point of view that people should be valued for their actions and behaviors and not their ancestries. I thought, 'So, we traced our ancestries—but we're all related any-way! Some of us are related and share some common patrilineal ancestry. Great—but we're all still a bunch of individuals.' I didn't see what we found as anything extraordinary. But that wasn't the way most of the world saw it. I was surprised by the extent of pub-lic interest in the Jewish, and especially in the non-Jewish, world."

"It was incredibly exciting to find something that could be trac-ing paternally inherited traits over three or four thousand years of history," Hammer said. "This is the first time ever we had been able to make a correlation with the ethnographic record over this timescale. It's a beautiful example of how father-to-son transmis-sion of two things, one genetic and one cultural, gives you the same picture. Some people keep records that go back three, maybe four generations. But fifty generations!"

Remarkably, the biblical and genetic stories of Aaron's legacy were aligned more closely than anyone might have believed possible. "The fascination, I think, stems from the biblical connection," Skorecki added. "It's like an archaeological find. When I thought about it, if someone were to have dug in the Sinai and found some of the original anointing remnants or relics of the original priesthood, I would have gotten very excited. This is a bit akin to that. I guess I just didn't appreciate how powerful a tool it was. I have to give credit to Neil and to Mike for saying, 'This is really, really very interesting.'"

The findings struck a nerve within the huge and growing worldwide community of amateur Jewish genealogists and other Jews curious to piece together the evidence of their ancestry erased during the diaspora or Holocaust. Like adoptees in search of their birth parents, Jewish men besieged the researchers with calls and letters, eager to find out if their DNA might "prove" their priestly inheritance. "I have been inundated with requests for testing," Skorecki said at the time.

The interest was, of course, over the top. The data were too fuzzy at this point to definitively prove much of anything. Yes, they had identified differences between most Cohanim and other Jews, but the number of mutations they looked for and the number of Jews tested were small. And according to Jewish law, oral tradition and not genetics is the one and only measure of Cohanim status. But that didn't deter hundreds of wannabe Cohanim. "It's emotionally very charged," said Hammer. "I think maybe that the last couple of generations have lost touch with their religious sources. They hear they're on this lineage, but it's not part of their life. Now they want to know if they really truly are."

Amidst the excitement of the discovery, a back current of controversy stirred, set in motion by the fervid public reaction and some sloppy journalism. In their letter to *Nature*, the researchers had been appropriately circumspect in framing their findings. "According to Jewish tradition," their article begins, underscoring their belief that the biblical story of the Cohanim is a tradition and not uncontested history. As they wrote, the date for the appearance of the haplotype could not even be traced definitively to the supposed time of the Exodus, let alone to Aaron. And as of yet, the evidence was not even strong enough to conclude that the common Cohanim mutation was a "founding modal haplotype" from a single individual.

However, in an interview with the *New York Times*, Skorecki, understandably enthralled by the findings, had enthused, "The simplest, most straightforward explanation is that these men have the Y chromosome of Aaron." Standing alone, the quotation appeared to be an unfortunate choice of words. Despite the careful wording of their article, mutated interpretations of the Cohen study spread like viruses, widely and wildly, in the media and on the Internet. The usually reliable *Science News* headlined its story "The Priests' Chromosome? DNA Analysis Supports the Biblical Story of the Jewish Priesthood," a gross overstatement. The Jewish Telegraphic Agency, which operates as a kind of Associated Press for Jewish publications around the world, reported breathlessly, and erroneously, that scientists had found that "the Jewish priestly lineage can be genetically traced back to the progenitor of all Cohanim, the biblical Aaron."

It was a public-relations mess, and science alone could not resolve the story of Aaron and his descendants.

AARON'S TOMB

History offers little clarity. The belief in a secure lineage of Jews that traces back to the Exodus and Aaron rests on shaky historical ground shaped almost entirely by myth and the unverifiable accounts in the Bible. The story of the first Hebrew priest is now part of the Jordanian legend of the fabled city of Petra in the southern Jordanian desert. Walid, my traveling companion, took me to Wadi Musa, the Valley of Moses, which bisects Petra near where, by tradition, Mount Hor is located and Aaron was buried. We checked in at the Taybet Zaman Hotel, a five-star resort, complete with its own private helicopter-landing pad, encompassing a renovated nineteenth-century village with antique style, traditional architecture, and period detail perched on a cliff overlooking the fabled city. Its $300-a-night rooms were going begging for $65.

Early the next morning, the brilliant sun sent shadows glancing down the rocky precipice and across the Arava Valley that snakes westward to Israel. Far below the hotel terrace, towering out of the rocky floor of the basin, cypresslike trees marked the hidden entrance to the city. From our vantage point on the edge of the cliffs, I could see a round, white dome framed in a yellowish halo appearing to float atop a distant mountain.

"That's Jebel Haroun," Walid said. Jordanians believe it is the Mount Hor of the Bible. "That building . . . that is Maqam Nebi Haroun, Aaron's tomb. We have been taught that Moses and his brother Aaron traveled to Petra and that Aaron was buried atop that mountain during the flight of the Hebrews out of Egypt."

Petra was founded by nomadic tribes who gave the city its Greek name, which means "rock." At the height of the Hellenistic period, in the second and first centuries BCE, the Nabateans had established Petra as a prosperous trading oasis on the camel routes between Gaza, Syria, and Arabia. Frankincense and spices, precious gems, ivory, and rich fabrics passed through there on the way to Greece, Rome, India, and Europe. The Nabatean Kingdom would demarcate the northern frontier of the Arab world until after the Muslim prophet Muhammad's death, in 632. Until that time, few Arabs lived in the Holy Land, much less in Jerusalem.

We entered the preserved ruins near where Moses and Aaron might have, through a narrow cleft in the astonishingly white rocks. The clip-clop of our donkey's hooves echoed in the still morning air as it passed through the winding 3-kilometer *siq* cut by time that leads to one of the world's archaeological marvels. The narrow fissure is believed to be responsible for the local legend that Moses struck the rock inside Petra to secure water for his flock during the Exodus. The path empties into an ancient courtyard, now eroded into sand and framed by sandstone walls etched by wind and rain with veins of red, purple, and pale yellow. Visitors are greeted by the massive Khaznet al-Faraoun, the Treasury of the Pharaohs, as it has erroneously come to be called, a soaring façade 100 feet high, a fascinating amalgam of Hellenistic and Middle Eastern elements. It inspired romantic tales of a "red-rose city half as old as Time," as rhapsodized by John William Burgon, an eighteenth-century Oxford biblical scholar (who had not visited Petra when he wrote this purplish prose).

What appear to be eroded, grand government buildings stretch out along a winding path through the city, but what survive today are actually the remnants of a necropolis. They were once royal mortuaries with magnificent façades to distinguish them from the plebeian burial sites carved in rocks in the less-fashionable sections of Petra. There are some eight hundred tombs, but only a few dozen with exteriors as elaborate as the Treasury.

Travelers in the nineteenth century spread the mistaken story that Petra was the ancient capital of Edom, which came to ruin because it denied shelter to the Jews on their flight from Egypt. Some say the queen of Sheba stored her treasures there en route to her tryst with King Solomon. Unaware of the city's ancient history, the Bedouin once believed it was the work of black magic. According to local legend, the Egyptian pharaoh, in pursuit of Moses, Aaron, and the clan of Hebrews, reached Petra after his slight detour at the Sea of Reeds. But upon arriving, the weight of the gold and jewels that he had thoughtfully carried along became too great a burden for his army. He created the Khaznet and deposited his riches in the urn at the very top of the façade, out of human reach. The detail of the Treasury's face is so exquisitely preserved that it looks too perfect, almost like a Hollywood stage set. (It should come as hardly a shock that it would take Harrison Ford, in *Indiana Jones and the Last Crusade*, which was filmed in part in Petra, to release the royal plunder.)

During my visit on a bright November day, only a few dozen visitors padded past the acres of soaring façades. Walid negotiated two camels from a Bedouin named Ibrahim, who had been posing with the animals for tourist pictures in the courtyard in front of the Treasury. Ibrahim is a Bdul, an ancient tribe and one of Jordan's poorest. Unlike most Bedouin, who can trace their lineage back to a founding father, the Bdul's heritage is a mystery. Some tribesmen believe they are descendants of the ancient Israelites or one of the breakaway Arab branches. Ibrahim led our three-camel caravan up the back rump of the mountain, toward Aaron's tomb.

Entirely covering the bald peak of the mountain, 300 feet above sea level, stands Nebi Haroun, an unremarkable, whitewashed, domed building with a haunting hold on three great religions. The original building was built in 1459, probably by Christians, who, legend has it, believed that it was where Aaron and the Israelites waited for Moses to return from Mount Sinai. Muslims revere this place because of its links to both Aaron and the prophet Muhammad. In the sixth century, Muhammad, then ten years old, supposedly climbed the mountain while passing through with his uncle on his way from Mecca to Damascus. There he encountered a Greek Christian monk overseeing the building, who prophesied that he would change the world. According to Ibrahim, Jewish visitors in the eighteenth and nineteenth centuries engraved Hebrew inscriptions on the tombstone house in an interior chamber as homage to Aaron. "The Jews go there to prostrate themselves and pray over his tomb, and nobody prevents them from doing so," wrote an anonymous Jew in 1537 in *The Genealogy of the Patriarch and the Prophets.*

"I know that it was not just chance that Muhammad came through Petra," Ibrahim said proudly as we clambered onto the roof. "He is from Aaron and Moses." Muhammad was drawn to Mount Haroun, he said, because it is the biblical Mount Hor "on the boundary of the land of Edom," where Aaron is taken to be "gathered unto the dead." According to the story in Numbers 20:23–29, it was the fortieth and last year of the wanderings of the Hebrews, and they were finally on the frontiers of the Promised Land. God has decided that Aaron should be stripped of his priestly vestments for having sinned years earlier in the sojourn by helping the Israelites mold a bull or calf out of melted gold—the famous Golden Calf—an idol much like El, the Canaanite high god. Moses is ordered to strip Aaron of his priestly vestments and put them on Aaron's son Eleazar. The first Hebrew high priest then dies on the summit, his sons left to carry on the priestly tradition.

Is the story of Aaron and his emerging role as a defining figure in Jewish identity real or myth? That's what a worldwide team of geneticists was on the verge of determining.

A MARKER OF JEWISH IDENTITY?

The genetic sleuths in Arizona, London, and Haifa had been stung by the criticism of their first Cohanim study. They believed they had been careful in evaluating the DNA samples. They were not trying to prove the story of the Bible but only exploring whether there was genetic witness to the tradition of the Cohanim. To do that they needed both more data and more brainpower. "We realized that more genetic markers could not only strengthen the conclusion, but allow us to do a better dating of the priesthood's origin," said Skorecki.

By this time, Bradman had decided that he wanted to launch his own laboratory. He had already discussed the limitations of the original Cohen study with a bright young geneticist, Mark Thomas, who was working at the Department of Biological Anthropology in Cambridge. Thomas had earned his doctorate in genetics at the University of Liverpool, where Bradman took his master's. The two had immediately hit it off, as they shared an interest in the application of DNA to historical questions. Bradman set his sights on luring Thomas to anchor his new laboratory, and he used the open-ended project on Jewish ancestry as bait.

"With their claims of common ancestry and their unique place in history, Jews struck me as a very interesting group," said Thomas, a lean, handsome man with the confident air of a striker on a championship soccer team. He jumped at Bradman's invitation. Bradman funded the lab while Thomas came up with the name: the Center for Genetic Anthropology. "Neil was asking fascinating questions. But in order to expand on the study, to attach a more accurate date to the origins of the lineage, they needed more information, more genetic markers on the Y chromosome."

Bradman also recruited David Goldstein, an American-born, Stanford University–trained geneticist who had moved to England in the late 1980s and quickly emerged as an international star because of his pioneering work on microsatellite markers. Goldstein was one of the first geneticists to use them to identify disease susceptibility genes by using what is called linkage analysis, which looks for markers that appear close together on a chromosome and are therefore probably inherited together. These blocks can shed light on all kinds of questions, such as population size, genetic variation, and when people migrated and intermixed. Most critical for Bradman, Goldstein appreciated the potential of using microsatellites for kinship studies.

"I recall getting an e-mail from him suggesting that we might get together for a meeting or maybe even dinner at his house," said Goldstein, who has since taken a position as director of the Institute for Genome Sciences and Policy at Duke University. Although struck by Bradman's friendliness, he nonetheless almost declined to see him. "I was taken aback that the signature below his e-mail message read 'Neil Bradman of the Jews.' I thought maybe he was an utter lunatic. It turned out that it was just from cutting and pasting another e-mail. When we met, we hit it off instantly. It was quite clear that there was a lot complementary in our views of how to do genetic history."

Like Hammer, Goldstein was raised as a secular Reform Jew. "I was never very religious. But as far back as high school, the history of the Jews has been of particular interest to me. It is a striking history, and it touches on all of the big features of Western culture." Early in his career, he had actually mused about doing a project on the genetics of Jews, but he never pursued it because he didn't think he would find much of interest on the Y chromosome. "Then I heard about the Cohen project that Neil and his son had worked on. I had barely heard about Cohanim. Maybe I had heard a vague reference once. But their project was interesting. They had figured out a way to address Jewish history. It immediately seemed to be worth doing. And I was taken by Neil's enthusiasm."

With the expanded London team in place, the identity seekers collected additional samples until they had tested more than three hundred male Jews, one-third of them self-identified Cohanim. Test tubes filled with DNA were sent off to Tucson. When the results for all the Jews tested came back, the Y team was stunned. Those Jews who did not consider themselves Cohanim had a variety of genetic markers on their male chromosomes, none of which stood out. In other words, as might be expected, the vast majority of Jews looked genetically quite diverse. In contrast, more than half of those claiming to be descendants of Aaron shared identical markers—not just one marker, but a cluster of six that the researchers dubbed the Cohen Modal Haplotype (CMH), with "modal" meaning "most common." The researchers estimated that the chance of these findings' happening at random is smaller than one in ten thousand!

While the overwhelming majority of today's Jews have haplotypes showing their combination of many different paternal lineages, the majority of Cohanim have just one—the CMH. "What we found was that approximately 50 percent of the Cohanim belonged to one haplotype, and many more were only one mutation distant," said Skorecki. If the researchers included a variation on one microsatellite, the similarity soared to well over 60 percent. The genetic marker was found in less than 10 percent of the general Jewish population. Why would the Cohen marker show up at all in common Jews? Some descendants of priests could have lost touch with their oral history and merged into the Israelite line. Goldstein, for example, has no knowledge of being a Cohen, yet he has tested positive for the CMH. If the Aaronite theory were to hold true, a tiny spillover of the marker into the general Jewish population (and to gentiles) would be expected, traceable to Jews who, like Goldstein, may not know of their ancestral past. Genetically, if not by Jewish tradition, they might still be Jewish priests.

The CMH is found in the J haplogroup, one of the streamlined groupings of eighteen major lineages, which geneticists letter A through R. Subgroups, or clades, of two haplogroups—E and J— are disproportionately common in Ashkenazi Jews and are colloquially described as "Semitic" or "Jewish," though because of their ancient origins, they show up in other populations. It's believed that the man who originated the J lineage lived in the northern part of the Fertile Crescent thousands of years before some of his descendants migrated to the Middle East 7,500 years ago. It is found at its highest variety in the Zagros Mountains in western Iran and in Iraq, where 60 percent of the population test positive for it. One branch of J, designated by geneticists as J1, is restricted almost exclusively to Middle Eastern populations, and this is where the CMH marker is most commonly found. Another clade, J2, which also includes Ashkenazi Jews, is also common throughout the Mediterranean countries and into India.

At least since the destruction of the Second Temple, the Cohanim have been nothing if not faithful. Based on their high rate of genetic similarity, Jewish priests have the highest "paternity certainty" rate ever recorded in a population genetics study. The CMH may symbolize religious and marital fidelity, but over how many centuries back in time? Unless it could be precisely dated, the ancestral line it represented could have originated around the time Aaron was supposed to have lived, or centuries before or after. That's where Goldstein's expertise proved invaluable.

"I spent a weekend at Oxford playing around with the developing manuscript," he recalled. "And it occurred to me that I could do the dating in a new way. You start with the assumption that an ancestral haplotype is the common one. Then you count the differences marked by the mutations back in time, but make the statistical correction for some of the mutations that are not visible." Although this seems straightforward and perhaps even obvious now, it had not been done this way before.

Bradman had the data with him in Israel, so Goldstein rang him up. "He had it all summarized, so he began feeding me the data over the phone. I was making the calculations as he talked, in my head. And when he read me the last batch, I could see where it was headed, how many years back the mutation probably originated, and I just stopped. 'Neil, do you know where this is going? This is going to three thousand years ago!' We just went silent on the phone."

Goldstein, not prone to overstatement, describes it as a Eureka! moment. "Neither of us was thinking, 'We've traced this chromosome to the time of Solomon's Temple!' Neither one of us was thinking anything of this sort. But we realized that it did mean something. It was a thrill, and definitely a sense of accomplishment."

Goldstein then ran the numbers that had been swirling in his head through a computer. After factoring in the uncertainties about mutation rates, he came up with a coalescence time of 106 generations ago. With a generation equal to twenty-five to thirty years, that would date the appearance of the CMH to 2,650 to 3,180 years ago. Perhaps the person who carried the haplotype—and who may have founded the Jewish patrilineal dynasty—lived before the destruction of the First Temple in 586 BCE, and conceivably further back in time, when the Exodus might have occurred.

It's easy to misinterpret what the CMH might be telling us. "I don't maintain that an approximate date of three thousand years ago actually confirms the story of Aaron," Goldstein told me. "That's a religious, not a genetic conclusion. It's a romantic similarity, though. There were priests taken away by the Babylonians when the Temple was destroyed. So, one view might be, although the priests dispersed, there was a common origin. If that pattern holds up, we can use [the CMH] as a signature of some type of ethnic association with Jewish populations."

The CMH has also been identified in some non-Jewish populations, but the frequency is low, nowhere near the high percentage in self-proclaimed Jewish priests. Each of those populations—Kurds, Armenians, Hungarians, and southern and central Italians—are believed to share some common ancestry with Jews, to have extensively intermingled with them, or to be descendants of converts out of Judaism.

One thing is clear: the CMH cannot definitively prove the existence of a single founding father for the Jewish priesthood, let alone confirm that he was Aaron. If it is primarily a marker of priestly inheritance, why would it show up on two J lineages—most commonly on J1 but also on J2—that split thousands of years, maybe more than ten thousand years, before the time of Aaron? Moreover, some Jews with an oral history of being a Cohanim and no known record of conversion have neither a J1 nor J2 lineage. They are from the haplogroup E3b, which has Middle Eastern origins, or from R1b, which is common among Europeans and some Turks. How could that be?

One biblically based theory revolves around the dual lineages of the Jewish priesthood. Although the founding priestly lineage rests with Aaron, a second line emerged during the reign of King David. This line was founded by Zadok, who historians believe was probably not an Aaronite. The sons of Zadok were in ascendance from the time of Solomon until the Babylonian exile; subsequently both the Zadokites and Aaronites held positions in the Temple. It's also possible that a rival priesthood was preserved among the Samaritans, the people formed out of the remains of the devastated northern kingdom, with some priests from this "foreign" lineage ultimately blending back in with the Judeans. In the Second Temple period, the Hasmoneans also claimed priestly status but their ancestral "purity" has always been suspect. Additional lineages could have originated outside of the oral tradition by someone who bought or assumed high priestships. According to Flavius Josephus, for a time the office of Cohen Gadol was thought to have degenerated into a purchased position. Geneticists are now testing these various Cohanim-origin theories.

THE POLITICS OF GENETICS

After the controversy that greeted the first Cohen study, the research team took extra caution in explaining their more refined understanding of just what the CMH might tell us about the history of the Israelites. Skorecki repeatedly reminded interviewers, including me, that the haplotype is found in some non-Jews. Bradman and Thomas struck a similar tone in a popular article written for *Judaism Today*. "Notwithstanding the identification of the CMH, it is not possible to say that those are the markers of a 'true' Cohen or whether, indeed, there was a 'first Cohen'—be it Aaron or someone else," they wrote. "In a similar way, there is no Jewish haplotype and genetics cannot 'prove' whether someone is a Jew; that is a matter for religious authorities. Nor can genetics decide whether a particular community is or is not Jewish. What may be possible is to demonstrate either movement between or a common origin for two or more communities, which may be known from other data to qualify for the epithet *Jewish* or as an ancient progenitor of such communities."

Despite these cautious comments, a small but determined contingent of scientists remained unrelentingly hostile to talk of an "ethnic" DNA marker, Jewish or otherwise. "The suggestion that the 'Cohen Modal Haplotype' is a signature haplotype for the ancient Hebrew population is . . . not supported by data from other populations," wrote Avshalom Zoossmann-Diskin, a former student of Bonné-Tamir. He noted that a number of studies have since found the CMH among Iraqi Kurds, Armenians, southern and central Italians, Hungarians in the Budapest region, and Palestinian Arabs, although in lower frequencies than in Jews. "It's not a coincidence that the haplotype was named the 'Cohen Modal Haplotype,'" Zoossmann-Diskin added in a conversation at his home in suburban Tel Aviv. "It could have been called a 'Semitic haplotype' or even a 'Kurd haplotype.' Why choose 'Cohen'? It serves the political interests of those who want Israel to remain a Jewish rather than a secular state."

Zoossmann-Diskin is certainly correct that by naming the marker "Cohen," the scientists embroiled the finding in politics. But it is more innocent than he is making it out to be. It's not unusual for geneticists to name modal haplotypes after their primary population. There is a Dinaric Modal Haplotype representing a signature marker found first in Croats, but also in other populations; a Muslim Kurd Modal Haplotype that's also found in other groups, including Arabs and Jews; and signature haplotypes for Native Americans that are not exclusive to them. Moreover, the finding of the J1 CMH in other ethnic groups does not disprove that it's an ancient Jewish marker. Far more likely, based on migration patterns and other DNA data, pocket populations in Italy, Hungary, Armenia, and in Kurdish lands are of Semitic, and possibly of Jewish, descent.

The power of a modal haplotype, scientists remind us, is not that it can prove ancestry but that it can set to rest unsubstantiated claims of descent. It is protection against false witness. While other populations that carry the CMH, such as the Kurds, are ethnically distinct and geographically isolated, only Jews have a set of closely linked genetic markers found in ethnic communities in all corners of the globe. That's remarkable—an almost incomprehensible genetic confirmation of religious continuity and solidarity.

Even more heated criticism appeared in the United States, led by a controversial University of North Carolina–Charlotte anthropologist. "In most scholarly contexts at the turn of the twenty-first century, one might be quite stunned to see a scientific study that begins by assuming the literal truth of the Bible," charged Jonathan Marks, notorious for his acid fulminations against what he calls "folk race research" of Jews, blacks, and indigenous cultures. He trashed the studies as "pseudoscience" and "Mickey Mouse biology" and dismissed the Cohen researchers—Jew, Christian, Muslim, and Hindu alike—as "pious" and "credulous," all but accusing them of cooking the books to support their religious biases.

Few geneticists shared Marks's jaundiced views, and most believed he grossly mischaracterized the Cohen research. "Unfortunately, population genetics is one of those areas where people feel entitled to an opinion without actually bothering to really look into the details," Goldstein commented. "If Marks were more of a population geneticist, he would realize we've made very modest claims. I think it's patronizing in the extreme to say that this kind of research shouldn't be done because it may be misused. There are enormous benefits in studying genetic variation."

The Cohen studies were truly remarkable. Although the extended family of Jews shares a signature of its distant ancestry, Jews are genetically (and visibly) different. The Cohanim researchers were not looking for genetic similarity among Jews living near each other or with the same last names; they were attempting to see if Jews scattered around the world with a shared oral tradition—Indian Jews from Mumbai, black Jews from Johannesburg, Ashkenazi Jews from New York, and Sephardic Jews from Israel—had a common male ancestor, and approximately when he might have lived. Two people can share a Y (or mtDNA) marker but show little genetic similarity because their common ancestor was so distant and their overall genome has reshuffled many times over the centuries.

Mark Thomas, the lead researcher and author of the second Cohen study, shrugged off the accusations of a Jewish conspiracy. "Well, I'm neither Jewish nor religious," he said. Descendants of a founding father—more than likely a Canaanite and conceivably an Israelite priest—were linked not by last name but by their DNA, which confirmed an oral tradition that predates the first use of surnames. "Personally, I have nothing but friendly contempt for religion. Clearly, I don't have an agenda to demonstrate biblical stories to be true. I think the criticism is rather silly. There are claims of common ancestry; there are origin myths for populations the world over. It's interesting to explore those, to separate truth from fiction."

The discovery of the Cohanim marker, while short of proving the biblical story of Moses and Aaron, does align with biblical history, oral tradition, and the archaeological evidence. The oldest original biblical text ever found is believed to be the Bircas Cohanim. The Israel Museum in Jerusalem has on display two small silver scrolls inscribed with a Hebrew prayer found near the Old City in the area of burial caves believed to be from the First Temple period. Whatever uncertainties may exist in the dating of the CMH, the DNA is testament to the certain relationship between the Jews and their God.

These more definitive findings also have had particular resonance among fundamentalist Christians and righteous Orthodox Jews, and for much the same reason: they endorsed, however faintly, prophecies in the Bible about the coming—or Second Coming, according to Christians—of a Messiah. "This is powerful," said Rabbi Ya'akov Kleiman, an American Jew who moved to Israel twenty years ago and now runs the Center for Cohanim in Jerusalem, which he founded. "God keeps His promises that we wouldn't remain scattered. The end is the redemption, and part of it is a functioning Temple. We see now that the exile is ending."

Like many North American Jews, young Jacob grew up disaffected, "a bagels and lox kind of Jew," he joked about himself as a teenager. Taken by what he calls "spiritual Zionism," he decided during a trip to Israel to remain in the country and later studied for the rabbinate. He changed his life around, including his name, from Jacob to Ya'akov. Unlike in North America, where most rabbis head congregations, there are thousands of rabbis in Israel who devote most of their time to Torah study and educational pursuits. For Kleiman, the Cohen finding was a sign from God.

"We're not messianists and we're not pushing the train, but it's going. This gene shows that. It shows the lineage of the Cohanim is true," he told me. With the redemption in mind, Kleiman is working through the Internet to track down Aaronites around the world, a formal listing of all known Cohanim. "Such a register existed in the days of the Sanhedrin," he added with evident pride.

FINDING DAVID?

The publicity storm over the Cohen lineage even spawned a feck-less attempt to identify what may be the most exalted biblical fam-ily ancestry, the House of David. The Hebrew Bible records that God promises that David and his descendants will always rule the Kingdom of Israel. "See, a time is coming—declares the Lord—when I will raise up a true branch of David's line," reads Jeremiah 23:5–6. "He shall reign as king and shall prosper, and he shall do what is just and right in the land." According to 1 Chronicles 3, King David had twenty-two sons, one of whom died in infancy, and one daughter.

Davidic lineage is of course a sensitive issue for Christians. It is stated in the Christian Gospels that Jesus was descended from David and is thereby the lawful as well as spiritual king of Israel. But the genealogical story in the New Testament is muddy. It lists two dif-ferent genealogies for Jesus, one in Matthew and one in Luke. Many Christian biblical scholars maintain Matthew is stating Joseph's line and Luke is stating Mary's line, which supposedly stretches back to Nathan, one of David's sons. The Joseph lineage is problematic, however. Although Jesus's earthly father was Joseph, who raised him, the Bible says Jesus was not of his seed, but the Son of God. Jesus's resurrection was therefore necessary to fulfill Jeremiah's prophecy that he is divinely descended from King David.

Over the centuries, both Jews and Christians have invoked descent from the royal house of Israel. Many European monarchies have claimed Davidic origins. The English royal family, the Windsors now ruling England, even have a rock from the Temple Mount under their coronation seat to emphasize that link. Jewish lore mentions dozens of lineages supposedly tracing to King David. Some great rabbis have claimed or been ascribed royal descent. The editor of the Mishnah, Rabbi Judah ha-Nasi, was considered to be of the Davidic line. His heirs, including the influential eleventh-century Jewish scholar Rabbi Solomon ben Isaac, known as Rashi, of Troyes, France, by tradition is considered a descendant of King David. Some rabbinic sources also claim Davidic lineage can be traced from Rabbi Judah Lowe, the sixteenth-century creator of the legendary golem, the giant statue of clay that he fashioned to protect the Jews of Prague from persecution by Christian religious authorities. Indubitable genealogical proof of Davidic ancestry is scant, however.

Could DNA answer this question?

Working with Family Tree DNA, Susan Roth, author of *Moses in the Twentieth Century*, has begun collecting genetic samples from various rabbinic surname lines in hopes of echoing the dramatic findings of the Cohanim studies. But most geneticists look askance at this genealogical escapade. Those who claim ancestry back to Rashi face the ticklish problem that he had no sons. Equally far-fetched is establishing genealogical continuity from biblical Israel to the Middle Ages to modern times. That would require passing through the gauntlet of Jewish Ashkenazi history, which includes dozens of pogroms, the mass slaughters of the Crusades, the Black Death, the Thirty Years' War, and the Holocaust, among other catastrophes. Researchers got lucky in identifying the Aaronite line. But with no credible genealogical evidence of a preserved line tracing back to King David and no oral tradition as in the case of the Cohanim, geneticists say it's hopeless to try to identify a signature "Davidic haplotype."

Those facts don't deter the devoted or the obsessed, who see in DNA the hope of confirming biblical prophecy and emboldening Jewish identity. "Throughout history, civilizations rose and fell, yet the Jewish people survived," Roth says. "Now we are back in our country in Israel and Israel is still being persecuted. Israel is not just a state. In light of what Israel is going through, it is imperative to show that the blood of King David is alive and well in Israel today, and that WE are the rightful inheritors of Israel."

PART II

HISTORY

CHAPTER 6

THE CRYSTALLIZATION
OF JEWISHNESS

The Israelites first appear in the historical record in an inscription dating from the thirteenth century BCE commemorating a victory of the Egyptian pharaoh Merneptah. "Carved off is Ashqelon, seized upon is Gezer," reads the stela, an inscribed pillar. "Israel is laid waste, his seed is no more." But what is this Israel? Is it a city-state? Wandering nomadic Semites? Descendants of the biblical Jacob? It's not clear, and the archaeological evidence does not help much. The "Israel" mentioned in the inscriptions disappears into the shadows of history for more than three centuries.

The Hebrew Bible is not much help in charting the transformation of the Israelites into Jews. According to Jewish and Christian tradition, God dictated the five books of the Torah—Genesis, Exodus, Leviticus, Numbers, and Deuteronomy—to Moses on Mount Sinai around 1200 BCE, months after the Hebrews left Egypt for the "Promised Land." We know that's apocryphal.* The key bibli-

---

* It is logically impossible for Moses to have received all of the five books at Mount Sinai because the complete desert journey, the details of which are supposedly inscribed by God on tablets, had only just commenced. At most, if divine intervention indeed occurred, God revealed the 613 commandments (the most notable of which are the "ten words" of the covenant, the Ten Commandments).

cal stories were not composed as a single saga but as tribal stories sliced, diced, and stitched together over many centuries, often reflective of the fractious geopolitical realities of the times in which they were written. The full five books did not take final shape until the third or second century BCE, by which time Judaism as we know it today had begun taking shape.

There is absolutely no evidence that a distinct Israelite population numbering in the tens of thousands, let alone the hundreds of thousands as described in the Bible, was living in Canaan or had recently arrived from Egypt about the time the Israelites were believed to have arrived. There is no sign of a violent invasion or endless conflict with Canaanite idol-worshippers. In contrast to the biblical account, the tribes were probably of mostly local origin. The fertile highlands of the region appear to have been continuously populated by farmers, who produced wheat, wine, olive oil, and figs in great abundance. They subsisted by trading, mostly with Egypt. Perhaps a few wandering tribes from the Sinai migrated into Canaan, but there is no hard proof even of this.

Based on their reading of the ambiguous archaeological evidence, Israel Finkelstein and Neil Asher Silberman believe that the invasions of migrant marauders known as the Sea Peoples and the resulting chaos around the region had shattered the political stability and trading networks of the ancient Middle East. According to these controversial but respected scholars, itinerant Canaanite farmers may have been forced to settle into small communities to survive, perhaps joined by slaves who had abandoned their masters. More than likely, competing tribal leaders (who would come to be referred to in the Bible as "the judges") ruled Canaan in what approximated a state of anarchy. It's now widely believed that the Israelite kingdom arose out of a core group of Canaanites mixed with the Sea Peoples (later known as the Philistines), Amorites (western Semites, like the Canaanites), some Hittites (a non-Semitic people from Anatolia and northern Syria), Hurrians or Horites (also non-Semites who inhabited parts of Syria and Mesopotamia), and Amalekites (nomads from southern Transjordan).

"The emergence of Israel was an outcome of the collapse of the Canaanite culture, not its cause," wrote Finkelstein and Silberman. "The early Israelites were—irony of ironies—themselves originally Canaanites!"

Unknown to history, sometime around 1000 BCE, a people calling themselves Israelites emerged as a local power in Canaan. They shared a common history but were divided into twelve often fractious tribes, which originally consisted of Jacob's sons: Reuben, Simeon, Levi, Judah, Dan, Naphtali, Gad, Asher, Issachar, Zebulun, Benjamin, and Joseph. Levi's descendants, the Levites, were the priests who were scattered through all the tribes. Jacob subsequently replaced the tribe of Joseph with Ephraim and Menasseh, the descendants of two sons of Joseph by his Egyptian wife, Asenath.

The Bible says that the judge Samuel anoints Saul from the tiny tribe of Benjamin as the first king. Known for their fierce warriors, the Benjaminites were situated between the Judeans and the northern tribes. But the diplomatic choice of a tribal son of Benjamin did not work out as planned, and the mantle of power soon passed to the much larger tribe of Judah. During the reign of King Saul, a young shepherd from the tribe of Judah named David of Bethlehem helped repulse the threatening Philistines led by the giant warrior Goliath. He later ascended to the throne and smothered the remaining Canaanite resistance. After a stay in Hebron, King David anointed Jerusalem as the capital and new spiritual center for the Israelite people. According to the historical books of the Bible, mainly Kings and Chronicles, under David's and his son Solomon's leadership, the fledgling tribal confederation was transformed into a glorious and wealthy empire.

With the portable Tabernacle replaced by the sumptuous Temple, the Israelites believed they finally had secured their homeland. Literature flourished among the priests and educated elite. Scholars believe that during this period, the court history of David in the book of Samuel and the apocryphal origins of the Hebrew people in the Torah were written. However, the United Monarchy would not survive Solomon's death in 931 BCE. Despite his reputation as the wisest king of all, Solomon managed to alienate a good part of his empire. The priests in the countryside, accustomed to their independence and privileges, saw their influence eclipsed by the gleaming new Temple. According to 1 Kings 12, the masses, mostly poor, uneducated, and, for the most part, pagan, resented the powerful centralized aristocracy and the privileges showered on the tribe centered in Judea.

According to the biblical account, Solomon's only son, Rehoboam, exacerbated the tensions and proved himself an arrogant fool by threatening to unleash scorpions on the masses if they didn't fork over the high taxes he was demanding. Only the southern tribes of Judah and tiny Benjamin acquiesced to this tyranny. The remainder of the tribes seceded to create a new but politically unstable nation in the north, retaining the ancient name "Israel," establishing its capital at Shechem and building an official cult around the idol of Baal. Much like pre–Civil War America, these two communities would evolve along separate paths.

The worst fears of Israel began to come true. The Assyrian military juggernaut rolled southward from Damascus in 734 to 733 BCE, ultimately vanquishing the northern kingdom over the next decade. The king of Assyria deported the Israelites to what is today modern Syria and Iraq. In the biblical version of history, written by the surviving Judeans in the south, God let this happen because of the widespread sacrilege and idol worship in the north. "The Lord was incensed at Israel and He banished them from His presence," reads 2 Kings 17:18; "none was left but the tribe of Judah alone."

The Bible claims that as many as 120,000 Israelites were taken away. The Assyrians imported colonists to this reconstituted territory, which became known as Samaria. The Bible says that the peasants left in Samaria mixed with immigrants imported by the Assyrians to form the nucleus of a new mixed population. The Bible calls this amalgam the Samaritans, the name derived from *samerim*, which means "keeper of the law." According to the Bible, these Samaritans were no longer the pure descendants of the Hebrews, but forsaken by God because of their sins. With the Judeans

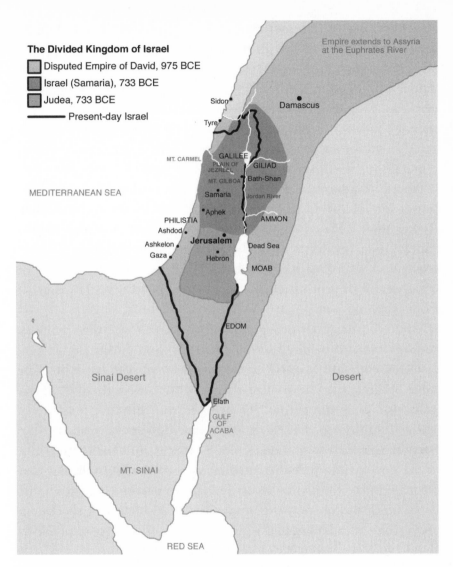

Figure 6.1. The Hebrews before the Babylonian Exile.

rewriting history, the ten tribes of the north thereafter vanish from the pages of the Bible.

What really happened?

The biblical account makes it seem as if the Judeans were spared because they were more pious than the strife-addled ten tribes of the northern monarchy. While the north is routinely dismissed as backward, pagan, and corrupt, Judea, with Jerusalem as its capital, is portrayed as the ardent promoter of monotheism and the proper heir to the glory of King David and King Solomon—the idealized model for Jewish nationhood. But that belief may be another case of biased propaganda written by and for the survivors.

Some historians and archaeologists believe that before the Assyrian invasion, a cosmopolitan nation with palatial government centers prospered across the verdant valleys and slopes of the northern kingdom. In contrast, Judea was rocky, sparse, and far less hospitable, with about one-tenth the population of the north. Rather than a Davidic golden age, as biblical lore holds, there is controversial evidence that the four-hundred-year-long dynasty controlled but a "marginal, isolated, rural region, with no signs of great wealth or centralized administration," and literally struggled for its very existence.

After the Assyrian occupation of the north, the ultrareligious Judeans came to believe that they had survived in this challenging environment only because of the zealousness of their faith. The Bible became a key tool in rewriting history, with the puritanical Yahweh cult as the heroes. It's believed that the first four books of the Torah, known as the priestly source, were burnished to reflect the territorial ambitions, cultic practices, and laws of sacrifice of these messianic Judeans. Introducing into the Bible themes that resonate today, the scribes recast Judaism as patriotic and grounded in the land, and uncompromisingly declared that a divinely chosen and living embodiment of the glorious reign of the House of David should rule over all of Judea and occupied Samaria.

This simplistic biblical account of pious southerners preserving the Hebrew tradition against the northern infidels is exaggerated and perhaps untrue. *The Annals of Khorsabad*, written by or on behalf of the victorious Sargon II, does not mention a huge influx of foreigners that supposedly corrupted the seed of the north. Rather than marrying immigrants to form a new "lesser" people, the sketchy archaeological evidence appears to show that many northern peasants and even some Israelite priests resettled in rural

Judea, as II Chronicles 30 hints. The biblical claim that the southern tribe of Judah, which gradually absorbed the tiny tribe of the
Benjaminites, remained "pure" and constituted the historical for-
bears of most of today's Jews while the northern Israelites were
tainted by foreign blood appears far-fetched. The southerners were
probably a hybrid of peoples, as were the northerners—and as was
almost every ethnic population of that region and time.

THE ORIGIN OF JEWISH RACIAL IDENTITY

Biblical scholars cite Deuteronomy as a turning point in the history
of the Hebrews. In writing their history, the surviving pastoral nomads of Judea combined fragments of the truth with liberal doses
of political propaganda. It was in part a manifesto of hope—a way
out of the cycle of triumph and destruction, piety and apostasy,
that had marked the prior centuries. This fledgling Bible would
provide a foundation for Judea's revival. Zealous Judeans strictly
followed the laws handed down at Sinai and established stringent
dictates against intermarriage.

The great prophets would dominate the next two centuries of
Jewish thought, offering an array of godly revelations. But Judea
was always at the center of the reworked narrative. Biblical historians believe that early versions of holy texts had Abraham settling
in the Valley of Jezreel in north-central Israel and his nephew Lot in
Transjordan. The new redacted version relocated Abraham's story
to the Judean hills. They emphasized the importance of the pious
Abraham, making him the central figure of Judean myth by placing him in Judea and in the center of a riveting narrative about the
unique origins of the people of Israel. Abraham's children were in
part the creation of political calculation! This near-final version of
the Bible concludes with the divine restoration of the Davidic Empire under pious King Josiah, who oversaw this round of biblical
writing and revisionism.

King Josiah, who ruled for more than thirty years, recovered some of the Israelite territories from the Assyrians, whose empire would mysteriously collapse over the following century. But still only a fraction of the size of modern Israel, the Judean nation remained vulnerable, its fate tied to his hold on power. In 609 BCE, with Assyria in sharp decline, Josiah was killed while trying to prevent an Egyptian advance into the region. At the cusp of the sixth century, tiny Judea found itself rudderless and in the crossfire between two great powers—Egypt, which had displaced Assyria as a power in the region, and the ascendant Babylon (the name is the Greek form of "Babel" from the Semitic "Babilu," meaning "The Gate of God") in Mesopotamia.

In the biblical literature, Babylon was synonymous with blood-thirsty wickedness and idolatry, although it was a sophisticated culture, famous for the magnificent Hanging Gardens. In 598 BCE, King Nebuchadnezzar laid siege to Jerusalem. The Judean resistance collapsed within months. According to 2 Kings 24:13–14, the Babylonian king carted away the treasures of the royal palace and the Temple and deported the elite to Babylon: "He exiled all of Jerusalem: all the commanders and all the warriors—ten thousand exiles—as well as the craftsmen and smiths; only the poorest people in the land were left." In 587 BCE, after suppressing a revolt, Nebuchadnezzar savaged the Temple and reduced the Jewish capital to a massive funeral pyre. Tradition has it that Judea was left an empty shell, its surviving population banished to Babylon.

The historical evidence of these banishments is much less certain. Based on recent excavations, the agricultural life in the countryside surrounding Jerusalem in the sixth century BCE remained relatively untouched by the Babylonians. It appears the victorious Babylonians deported at most a quarter of the Judean population, fifteen or twenty thousand people, numbers in line with the accounts in the books of Kings and Jeremiah.

Key portions of the biblical texts were significantly reedited right before and during the Babylonian Exile. Instead of criticizing the

hated Babylonians, many prominent Hebrew priests excoriated the
Judeans themselves. In the biblical version of history, it would fall
to the exiles of Babylon to rescue the Hebrews from the shattering
loss of their national identity. The expansive enclave prospered.
There is evidence of the Israelites as merchants, soldiers, civil ser-
vants, and the like, although Isaiah also makes reference to the
"poor and the needy." But Babylonian rule over Judea lasted only
a scant half-century before the Persian emperor Cyrus vanquished
Babylon in 539 BCE, and the distant territory came under his rule.
It was at this time that the Judeans began being referred to as Jews,
and Judea became known as Yehud (the Aramaic name of the new
province in the Persian Empire).

By this time, Judea was little more than a battered capital city,
Jerusalem, surrounded by a scattering of towns. Almost immediately upon assuming control over Yehud, Cyrus decreed that the
Temple should be rebuilt, and construction of the Second Temple
began in 516 BCE. Overjoyed, some hardy exiles, undeterred by
an unrelenting drought, commenced what would become known
as the return to Zion, although a sizable community of the most
prosperous Jews would remain behind in Babylon.

The still-incomplete Hebrew Bible had yet to be knitted into
the compelling account we have today. For the most part, Jewish
identity had been tribal and territorial, but that anchor had been
shattered. Certainly, if the central narrative of the Jews was to have
relevance for the majority of the Jewish people who then lived in
diaspora communities in Babylon and elsewhere, it could not be
left with the ephemeral restoration of the Davidic Empire under the
zealous Josiah and the limp return to a crippled homeland under
Persian rule. The Babylonian Jews, introspective and tempered by
years under foreign domination, determined that they had a divine
right to restore their dignity and their religion—to come up with a
more relevant and coherent story based not on their recent history
of failure but on the legacy of a glorious past.

Recognizing that a return to Zion could not be guaranteed, the
priests and scribes of Babylon, who were just beginning to commit

to papyrus what would become the Hebrew Bible, expanded the central focus of Judaism from God's promise of an actual homeland ruled by King David's descendants to the laws handed down on Mount Sinai. Two Jews, Ezra and Nehemiah, would prove key. They aggressively promoted a renewed covenant based on genealogy and common ancestry. Some rabbinical authorities go so far as to attribute to these prophets, or the scribes who worked in their names—their historicity is difficult to verify—the creation of Judaism or at least the notion of Jews as a "race." Many scholars today believe the claims of ethnic purity that appear in the Hebrew Bible as dated to the Exodus and the conquest of Canaan were inserted to bolster this new narrative. What had been minor references in early versions of Deuteronomy prohibiting intermarriage became far more prominent in these newer iterations, eventually hardening into Jewish dogma.

According to the Bible, Ezra returns to Jerusalem from exile around 458 BCE. Flush with funds provided by the Persian king, he assesses what needs to be done to restore the decimated Jewish homeland. He is shocked by the destitution he finds but even more by the flaccid moral lives of the Jews. They act like common Canaanites, Egyptians, and other gentiles, he says. The taboo against what geneticists call exogamy has apparently all but evaporated. The prototypical scolding parent, Ezra lambastes the Jews in Ezra 9 and 10 for their "abhorrent practices," most notably for intermingling their "holy seed . . . with the peoples of the land. . . . You have trespassed by bringing home foreign women, thus aggravating the guilt of Israel."

Intermarriage, even by the Levites and the Cohanim, is rampant, and high priestships reportedly could be bought, which if historically true, could be one reason why the Cohen Modal Haplotype shows up in more than one traditionally Jewish lineage. Fearful that the distinctiveness of the chosen people is being diluted, Ezra commands that the exiles now flooding into Jerusalem stick together. "So now, make confession to the Lord, God of your fathers, and do His will, and separate ourselves from the peoples of the land and from foreign women," he says. He produces a list of more than one hundred men who have renounced their foreign wives and children, establishing what amounts to a blood standard of Jewishness.

Ezra's rulings, along with the decrees of another returning Babylonian exile, Nehemiah, would burn the notion of Jewish separateness into Jewish identity. While Ezra of Babylon preaches spiritual redemption, Nehemiah of Persia stresses practical reforms. A high official in the court of the Persian king, Nehemiah presses his benefactor to fund the restoration of Jerusalem, which had yet to recover from the savage devastation of Nebuchadnezzar. In an echo of Ezra, Nehemiah criticizes those willing to marry outside the faith. "I censured them, cursed them, flogged them, tore out their hair, and adjured them by God, saying, 'You shall not give your daughters in marriage to their sons, or take any of their daughters for your sons or yourselves,'" reads Nehemiah 13:25.

In an apparent attempt to ensure that the new generation of children would be brought up Jewish, as defined by the male lineage, he decrees that Jewish men should divorce their foreign wives. Intriguingly, Nehemiah, like Ezra, does not address what if anything should be done about marriages between Israelite women and foreign men. While some scholars believe this is a deliberate seeding of what will grow into the matrilineal principle of Judaism, most others believe it says more about the chattel status of women in this era. Although the zealous Ezra and Nehemiah attacked intermarriage as a death knell for Judaism, their efforts were never directed at converts, just those who did not fully embrace the religion and its customs. Tactical decisions by exile scribes and priests forged the concept of the sacred purity of the "Jewish race." Over time, Jews began officially conceiving of themselves as a separate people, defined by absolute obeisance to Jewish law and marriage within one's own tribe: follow Scripture and the Hebrews will prevail as a people. Religious orthodoxy, the cult of Yahweh, and the ethic of Jewish exceptionalism and separateness had triumphed.

THE SAMARITANS

From today's historical vantage point, Abraham's covenant is represented in the ancestral line that extends from the House of David, through Judea, into the diaspora, and to modern Jewry. But the validity of that lineage appeared much murkier to the inhabitants of the Levant in the sixth and fifth centuries BCE. At that time in history, there were still two vital factions vying for the legacy of the ancient Hebrews—the Judeans returning to the decimated southern kingdom and the Samaritans to the north, numbering more than 1 million, who considered themselves descendants of the Israelites.

Providentially isolated, the priests of Judea had managed to sidestep the Assyrian conquest and the colonization inflicted upon the north. But renewed claims by Judeans that they had remained ethnically purer than their neighbors to the north were once again almost certainly overblown. If the accounts of the Hebrew prophets are fair measure, marital mixing was also the reality for the tens of thousands of Judeans who remained in the south during the Babylonian Exile. The Jews of Judea and the other large Jewish community of the time, in Egypt, often married non-Jews. While the peasantry did not wander far from their roots and married within their local communities, the aristocracy, including the priestly families, intermingled with imported foreigners. It was exogamy by necessity, and it didn't change much, at least at first, even after Ezra and Nehemiah supposedly issued their harsh dictates upon their return from Babylon.

The considerable number of Jews returning from exile considered themselves superior to the Samaritans, which only inflamed the ethnic cold war between the south and the north. It would persist for centuries. Jews who converged in Jerusalem after Cyrus's proclamation summarily rejected entreaties from the Samaritans to reconstitute the original land of the Israelites or help in rebuilding the Temple. The Samaritans in turn rejected what they considered an inaccurate oral tradition of Judea represented by the prophets—they were undoubtedly irked by the works of Ezra and Nehemiah—and the oral works later collected into the Mishnah and Talmud. They insisted upon the primacy of the original five books of Moses, a version of which had been written by this time, and stoutly resisted any further revisions to the text. Even today, the Samaritan version of the Torah is written in an ancient Semitic script. From this point onward, the Samaritans would be denied their biblical right to the Promised Land. Abraham's children gradually evolved into two different peoples: Israelite Jews and Israelite Samaritans.

Samaritan history, much like the story of the Jews, has been marked by a succession of massacres and misfortunes. The Samaritans did build their own cultic place of worship, on Mount Gerizim in ancient Shechem, which they considered the place of the original altar of the Hebrew god. Many early Israelites, and all Samaritans of the time, believed that most of the ancient prophecies and traditions regarding the "chosen place," including the sacrifice of Isaac and the prophecies of Moses, referred to Mount Gerizim, near present-day Nablus, and not the Temple Mount, in Jerusalem. The sanctuary was destroyed in 113 BCE, rebuilt, and then demolished again in the fifth century CE, after which it was restored as a Christian church. Muslims leveled it for good three centuries later. By the time of Jesus, the Samaritans, by then a dwindling presence, were widely considered outcasts. Traditional Jewry maintained that the Samaritans forever ruptured their biblical inheritance when they violated the laws of the Israelites and married outside the faith—a questionable claim, as history suggests.

From a nation of hundreds of thousands at its peak in the fifth century CE, the Samaritans gradually dwindled in number to about 150 at the end of World War I, and today number about 650, from four large families. Most live in Kiryat Luza, a town near Mount Gerizim, or in the town of Holon, near Tel Aviv. Although many fiercely maintain their ethnic identity and fewer than two dozen Samaritans have voluntarily dropped out of the community since the 1930s, for the most part they are well integrated into the Israeli community. They still believe they have followed the teachings of Moses more closely than Jews and are proud to have maintained the ancient pledge of their people to live in the Holy Land—and marry only within the community. The genetic insularity of the Samaritans makes them, like Jews, a favorite ethnic group of population geneticists.

Batsheva Bonné-Tamir knows more about the genetics of the Samaritans than any other scientist in the world. Her interest dates to the late 1950s, when she received a grant to study for a master's in cultural anthropology at the University of Chicago. "The students were from the United States, and they were mostly interested in studying American Indians," she said. "That didn't interest me. I went to the Hillel House on campus and found a book on the Samaritans. They are the most ancient, inbred civilization in the world. I was hooked right away."

Bonné-Tamir began regular treks to the two Samaritan communities, taking oral histories and swabs of DNA to reconstruct the community's evolution. She matched up genetic markers against family genealogical records, which the Samaritans have maintained faithfully for more than thirteen generations. She estimates that more than 80 percent of Samaritan marriages have involved either first or second cousins. Even more than Jews and over a far longer time, the Samaritan families represent the world's most homogenous lineage.

Genetic genealogy is also resolving the ancient biblical mystery about the true origin of the Samaritans. Historians have long been skeptical of the biblical account in 2 Kings 17:24 that the Samaritans are resettled foreigners, suggesting that scribes may have portrayed them as Assyrian interlopers to justify the rejection visited upon the prickly Samaritans by those Israelites returning from Babylonian Exile in 520 BCE.

What does the DNA say?

Bonné-Tamir has found that all Samaritan men are more similar to Jews than they are to their Palestinian neighbors. She has identified four distinct Samaritan paternal lineages: the Tsedakas, which tradition holds are descended from the tribe of Menasseh; the Joshua-Marhiv and Danfi lineages, which are supposedly linked to Ephraim; and the priestly Cohen lineage. Each of these male lineages can be tracked back in an unbroken line to the Assyrian invasion of the northern kingdom of Israel—and possibly to the founding of the House of David. The DNA of the women tells a different story, however. Samaritan females are no more like Jewish Middle Easterners than are Palestinians. In practical terms, that suggests that the progenitors of each of the four major Samaritan lineages took on local wives many centuries ago—perhaps Assyrian colonists.

Researchers in Israel and from Stanford University recently compared the DNA of the Samaritans who claim Cohen status with Ashkenazic and Sephardic Cohanim. Not surprisingly, and confirming Bonné-Tamir's work, the Samaritan Cohanim do not carry the ancient Cohen Modal Haplotype. But in a big surprise, the three non-Cohen Samaritan lineages shared the CMH cluster of markers or were very close.

What does that mean?

The scientists speculate that not only are today's Samaritans likely descended from the Israelites, they may be the ancestral remnants of a breakaway group of Jewish priests that did not go into exile when the Assyrians conquered the northern kingdom in 721 BCE. Instead, these Cohanim may well have stayed, "but married Assyrian and female exiles relocated from other conquered lands, which was a typical Assyrian policy to obliterate national identities." It may just be that the tiny clan of Samaritans are a rare surviving branch of the ancient Israelites.

THE JEWISH STATE

After two centuries of Persian rule, the Greeks became the next in the series of distant empires to rule over the land of Canaan. In 334 BCE, twenty-two-year-old Alexander of Macedonia attacked

Persia, executing a plan originally conceived by his father, Philip, who had been assassinated a few years before. Within a decade of conquering the once-formidable Persian Empire, Alexander had earned the honorific "the Great." His thirst for conquest helped spread Greek culture to the far corners of the world. Following his death, in 323 BCE, Alexander's fragile eastern kingdom in the Levant was divided between his generals. Tiny Judea, which under Greek influence took on the name Palestine, was caught in a vise between the Ptolemy Dynasty in Egypt and the rival Seleucids, who ruled over Syria, Phoenicia, and the lands east to Persia.

By this time, Jewish diaspora outposts in Egypt, Syria, and Iraq eclipsed in size and influence the community of Jews based in Jerusalem. "Our heritage has passed to aliens, our homes to strangers," the Hebrew Bible says of this period, in Lamentations 5:2–6. "We have become orphans, fatherless." Throughout the Greek Empire, Jews gradually became known by their unique devotion to One God and their ancestral connection to Abraham rather than as citizens of a particular nation.

Greek rule did bring a measure of stability to the Mediterranean. After years of slow decline, the Jewish population began to grow again, although it absorbed many foreign elements. Under the Ptolemies, the tide of Greek culture, with its celebration of philosophy, literature, and art, spread through every aspect of Jewish life. To many Jews, this mix of Judaism and Greco-Roman ideals was not only modern but liberating, an anticipation of the liberalization that would transform Jewry in Europe during the Enlightenment. But traditional Jews viewed the secularization of Jewish life in much the same way as ultrareligious Jews would look askance at many "reforms" introduced into Judaism in the eighteenth, nineteenth, and twentieth centuries.

These tensions escalated after the Seleucids of Syria, now under the thumb of Rome, wrested control of Judea from the Ptolemies. Installed by Rome in 175 BCE, Antiochus IV "Epiphanes" (the Illustrious) moved to further secularize Judea. The Hellenized Jewish intelligentsia embraced—and many excused—his campaign as the price to pay for modernizing an outdated religion. This process accelerated after Antiochus's ally, Menelaus, took over as the Jewish high priest in 171 BCE. He raised taxes and, in a fateful move, replaced Mosaic Law with secular statutes. Believing they had wide public support, Antiochus and Menelaus converted the Jewish Temple into an ecumenical place of worship for all the local citizens, which meant adding a statue of the Olympian Zeus.

It quickly became apparent that the secularists had overplayed their hand. Many Jews considered these changes religious genocide, drawing support from the masses and transforming the religious dispute into a political revolt. Matthias Hasmon, an elderly priest from the Judean foothills, sparked the uprising when he murdered a local supporter of Menelaus. "Let everyone who is zealous for the law and supports the covenant come out with me!" he is quoted as saying in 1 Maccabees 2:27.

In reprisal, the Syrians slaughtered a group of pious rebel soldiers who refused to fight on the Sabbath. The remaining revolutionaries put Jewish survival ahead of religious purity and fought ferociously, seven days a week. They were led by one of the old priest's sons, Judas, whose relentless raids earned him the nickname "the Maccabee"—the hammer. Judas the Maccabee capped his successful campaign in 164 BCE with the capture of the Temple and the reestablishment of the cult of Yahweh, which stressed the direct role that God plays in preserving the covenants and the place of the Hebrews as the chosen people. Within a few years of the uprising, Judas the Maccabee established the holiday of Hanukkah to commemorate the victory and as a celebration of the restored Temple, purified of pagan symbols. It would take another two decades for the Maccabean successors to prevail for good and establish a hundred-year religious Hasmonean dynasty. Although Palestine nominally remained a province of the Seleucids and under the watchful eye of the Romans, for the first time since the Davidic Empire, the Hebrews, now called Jews, had an independent state.

The Hasmoneans restored the conservative religious establishment, which closed the Greek academies, replacing them with Bible schools that would become the hallmark of traditional Judaism for millennia. They moved to nominally unify the fractious Jews by reinforcing the notion of Jewish separateness. "If there is any man in Israel who wishes to give his daughter or sister to any man who is of the stock of the gentiles, he shall surely die," reads an account in the Book of Jubilees, an ancient Jewish religious book that dates to this period.

The Hasmoneans eventually fell victim to their fanaticism. According to accounts by Flavius Josephus and in the book of Maccabees, fancying themselves as empowered by divine decree to restore the Davidic Kingdom, the Hasmoneans built towering mausoleums similar to the edifices that have survived in Petra. They employed an army of mercenaries, leveling the Samaritan temple, annexing Petra and non-Jewish Idumea—roughly what is today the Negev—and invading the Decapolis, the ten Greek-speaking cities around Jordan and Syria. At each stop, they massacred or forcibly Judaized the local pagan peasant populations. Established as a living embodiment of Jewish fundamentalism, the Hasmonean campaign of realpolitik resulted in the absorption of more gentiles into Judaism than any Jewish government or social movement in history.

The merging of nationalism and religious triumphalism eventually sparked fierce debates over what would become an issue for the ages: Who is a Jew? Cults proliferated and competed. Although there is no extrabiblical record of their existence, the legalistic and rationalist Sadducees (or Zadokites) were thought to have aligned with the wealthy elite, rejecting the mysticism of the masses and their belief in an afterlife, which is not supported by the Torah. The Essenes, a nomadic communal faction of ascetics, rejected the Temple leadership as hopelessly corrupt. The Pharisees, whose name means "the separated ones," resisted the alliance of convenience between Jewish politicians and the high priests and rejected the notion of a Jewish state, insisting that faith should be at the center of Jewish life.*

Pompey's occupation in 63 BCE brought Judea under direct Roman rule that would last for seven centuries. Hoping to avoid the mistakes of the Greeks, Rome resisted officially annexing the prickly Jewish nation, whose citizens had shown their willingness to fanatically fight and die for independence. It installed a vassal state, naming the loyal Herod, an aristocrat whose family had converted to Judaism during the Hasmonean period, as the new Jewish king. That ended for good the Davidic lineage of Jewish royalty. Herod the Great, as he became known, oversaw a bloody reign of enlightened despotism. He restored Jewish hegemony over much of the original Israelite territory, instituting broad reforms, targeting banditry, and introducing more sophisticated farming techniques to help the rural poor. He went on a building spree, most famously, around 19 BCE, refurbishing and expanding the aging Second Temple in Jerusalem. During Herod's reign and soon after, the Jewish population throughout the Roman Empire soared to over 5 million, heights that would not be seen again until the nineteenth century.

But what Herod gave with one hand, he took with the other. He systematically liquidated the last vestiges of the Hasmonean family and murdered his own wife and two sons. He stacked his royal court with Jewish priestly families imported from Babylon and the Roman West. The court was notoriously corrupt. His public obeisance to the national religion did not impress the splinter sects, who were convinced that Jewish Jerusalem had become a wholly owned subsidiary of the Roman Empire. It was no surprise that after his death, in 4 CE, the independent Jewish province collapsed; the country was divided into four provinces, each under Caesar's

* Historians believe the Pharisees observed a loose collection of ritualistic oral traditions and were obsessed with genealogical purity, the blueprint of historical Judaism. They are the forebears of contemporary ultrareligious rabbinical Jews who have become a potent and growing political force in modern Israel.

thumb. It was the last time a Jew would rule over a united Palestine until the founding of the modern State of Israel in 1948.

JEWISH CHRISTIANS

With Jerusalem now considered a backwater outpost, Rome shuffled through a series of foreign-born governors. Taxes soared. Aristocratic Hellenized Jews bickered with dissident factions, including a sect led by Jesus that biblical scholars believe was heavily influenced by the mystic and pacifist Essenes, who probably represented a sizable percentage of the Jews in Palestine of the time. According to the Bible, Jesus declared that the kingdom of God on earth, so devoutly desired by Jews and later by Christians (though in a radically new conception), could not be accomplished unless his pronouncements were followed.

Today, the Jewish people stand out as unusual for having one of the few ethnic religions in the West. Yet until the ascension of Christianity, it was only one of many tribal religions that melded faith and ancestry. Jesus represented a dramatic challenge to the Mosaic traditions that had come to define Judaism. For most Jews of that era, faith was rooted in the dictates of the Bible and the ancestral ties reaching back almost two thousand years. Jesus proposes what appear to be only subtle alterations to Jews' compact with God. "Do not think I have come to abolish the law or the prophets," Matthew 5:17–18 quotes Jesus. "I have come not to abolish but to fulfill." But while they appear small, the proposed changes, as filtered through his apostles, are revolutionary. Jesus rejects the belief that faith should be mediated through the laws, which were open to the interpretation of what many Jews of the time considered corrupt—or at least corruptible—Jewish priests. He sees the laws not in inscrutable texts but in each person, as a reflection of his or her individual relationship with God. He explicitly rejects the Jewish concept of kinship.

While the imprint of Judaism could still be found in the genes,

Christianity came to be centered in the soul. It is a momentous fracturing of the tradition of tribal ancestry as the defining component of Jewishness. For the followers of Jesus, henceforth faith would take precedence over scripture and ancestry. In his letter to the Galatians, Paul is intent not just on propagating the faith among those who strictly practice Temple Judaism, but also on converting the gentile masses. When God promises Abraham that through the covenant his offspring will be blessed, Paul argues, he is not suggesting a literal seed, but rather the seeds of faith. As Paul writes in Galatians 3:29, "If you belong to Christ, then you are Abraham's offspring, heirs according to the promise." In other words, if you accept Christ as savior, you are an adopted child of God and one of the chosen people. The Apostles stress a new covenant, symbolized by the blood of Jesus. Jesus is quoted in John 6:54 as saying, "Those who eat my flesh and drink my blood have eternal life." Christian blood rituals came to be seen as an allegorical foreshadowing of the atoning death of Jesus through which blood offered spiritual salvation for all humankind.

For Jews, the blood connection was never symbolic but literal and evidence of an ancestral line that stretched to the beginning of human time. Blood rituals are painstakingly detailed throughout the Torah. To protect the Israelites from the plague while in bondage in Egypt, God instructs them to smear the blood of a lamb on the door and mantel as a visible sign of their faith, the origins of the modern Passover ceremony. While sharing God's commandments, Moses splashes blood over the faithful. Like other ancient religious groups, the Israelites incorporated blood in the form of animal sacrifices as a holy form of consecration, a ritual offering from priest to deity. In the Jewish religion, even the symbolic consumption of blood borders on paganism. "[Y]ou must not eat . . . flesh with its lifeblood in it," God commanded Noah in Genesis 9:5. Because blood symbolized life, ingesting it was considered an act of sacrilege by the Israelites. The rule was absolute.

The reinterpretation of Abraham's covenant as spiritual gradually takes hold among the rebellious sect that evolved into Christianity. God indeed chose the Jews but passed the sacred covenant to the adherents of a "reformed" Judaism: what evolved into Christianity. To continue to insist, as Jewish law does, that genealogy plays a critical role in how we define ourselves religiously would deny the teachings of Jesus. Or as Paul put it in Romans 2:28–29: "[A] person is not a Jew who is one outwardly, nor is true circumcision something external and physical. Rather, a person is a Jew who is one inwardly, and real circumcision is a matter of the heart—it is spiritual and not literal." In a fateful split with Judaism, the genealogical inheritance was henceforth conceived metaphorically, not literally.

The centuries before and after the time of Jesus were marked by intense religious foment. Even Jews took to proselytizing, unusual for a religion that has historically set up barriers to conversion. From about 150 BCE to 200 CE, Israelitish Judaizing colonies sprung up in Italy, Gaul, and along the coast of Asia and Africa. In one mass conversion well known to history, in the first century BCE, Queen Helena of the small kingdom of Adiabene in ancient Kurdistan converted her entire household and many commoners to Judaism.

The mass of Jews, Jewish Christians, and new Jewish converts were united in loathing the Roman occupying forces, who continued to practice pagan rituals. In 66 CE, in rebellion against another round of Roman taxes and attacks on the Temple, Jewish priests in Jerusalem stopped offering daily sacrifices on behalf of the emperor. The protest ballooned into a guerrilla revolt and then a full-fledged war. Messianic zealots, bandits, and freedom fighters banded together to fight a battle known in Roman history as the Jewish War and remembered by Jews as the Great Revolt.

Almost all the details of the conflict come from the pen of Flavius Josephus, whose account is viewed as contradictory and unreliable. As the tensions in Palestine escalated, the new Roman emperor Vespasian sent his son Titus to attack the city in 70 CE, which he did with unbridled zeal. After four years of sporadic but often vicious fighting, the nominally independent Jerusalem fell. The Temple was left in ruin, with only the Western Wall left standing. The sacking destroyed the Jewish genealogical records, meticulously assembled over hundreds of years. The last gasps of resistance collapsed around 74, when hundreds of zealots holed up in the fortress at Masada killed themselves rather than fall into the hands of the hated Romans. "Masada," as the mass suicide is referred to, has come to symbolize the unyielding commitment of Jews to preserve their beliefs and uniqueness even in the face of certain death.* By the time the clashes subsided, hundreds of thousands of the Jews of Palestine may have been killed. Tens of thousands fled or were sent away, many deported to Italy as slaves by Titus, adding another wave of exiles to a growing Jewish diaspora.

Over the coming decades, the Romans faced uprisings in Jewish communities in North Africa, Libya, Cyprus, and Alexandria, whose famous synagogue was reduced to rubble. Weary of the resistance and intent on placating Palestine to secure the eastern flank of the empire, Emperor Hadrian arrived in Jerusalem in 130 CE to assess the situation. Although he was initially sympathetic to the pleas of the Jews, who wanted more independence, he quickly turned hostile to what he concluded was a backward ethnic religion. Rather than rebuild the city as a Jewish capital as he had initially promised to do, he Romanized it even more. Much to the horror of Jews desperate to see the construction of a Third Temple, he dedicated the city to the god Jupiter and renamed it Aelia Capitolina. He forbade circumcision, the keeping of the Sabbath, and the making and keeping of a Jewish calendar, which was devastating, since the community was so devoted to ritualistic study and prayer tied to Jewish holidays. In short, Hadrian believed he was

---

* The Masada story, which many historians contend is apocryphal, has political resonance in Israel. While Israeli soldiers are today required to take an oath, proclaiming, "Masada shall not fall again," some Israelis see the story as a more fitting symbol of self-destructive extremism. They consider Israel's reluctance to grant territorial concessions and more freedom of movement to the Palestinians as a "Masada complex," a form of self-imposed suicide, because it will eventually provoke a Mideast version of the Alamo.

modernizing biblical Judaism, an effort that had some appeal to the Hellenized Jewish upper class.

But the campaign of Romanization fell victim to yet another uprising, ignited in 132 by the followers of the enigmatic Simeon ben Kosevah, later known as Bar-Kochba (Son of the Star). He was a charismatic Jewish fundamentalist who, like Jesus, claimed he was a messiah. He viewed the revolt not as a religious struggle but as a race war, a desperate attempt to preserve the Jewish people and the concept of chosenness. Understandably, most Jewish Christians and nonbelievers refused to join the revolt. Hadrian got military support from Antiochus Sidetes, the Seleucid ruler in Syria, who concluded that Jerusalem "should be destroyed and the Jewish people annihilated because they were the only people on earth who refused to associate with the rest of humanity." Embattled, Jews fled Jerusalem en masse to what they hoped was the safety of the Galilee, but to no avail. In 135, the Romans supposedly liquidated hundreds of thousands of Jewish rebels. While devastating to the Jews, the revolts also inflicted heavy casualties on the Romans.

Defined by ancestry and ritual, sure in the belief of their chosenness, Jews were one of the few ethnic groups to fiercely resist Roman rule and fight to the death to preserve their identity. It was a renewed signal to history that these determined monotheists would not be an easy people to assimilate. Surrounded by polytheistic cults and in competition for adherents with religions and the new cult of Christ, they stubbornly maintained their steadfast belief in One God and One People, blessed by divine grace. It was almost as if geographic fragmentation were the price Jews had to pay for their spiritual and ethnic cohesiveness. Jews were no longer a nation; they were now struggling to merely survive.

DIASPORA JEWRY

In the decades after Jesus's death, the trajectories of Judaism and the religion it birthed moved in starkly opposite directions. In the time of King Herod, the majority of Jews lived in Palestine, and more than three-quarters resided in the Middle East. Within just a few hundred years—after the crushing defeat of the two Jewish revolts and the desecration of Jerusalem—the figures would be reversed— all but a quarter of the Jews had fled the biblical homeland. By the end of the second century, Palestine had lost its signature identity as primarily Jewish. The Roman massacres during Hadrian's rule turned the mostly urban Jewish population, devastated and spiritually besieged, into poor farmers. Palestine was left a demographic garble of Samaritans, a sizable community of Hellenized pagans, a growing number of Christian converts, and a dwindling population of Jews.

While the cult of Jesus gained adherents, a steady stream of Jews left the faith or disembarked for distant lands. Although population estimates under Roman rule are only speculation, it is believed that the Jewish community in Palestine stabilized at about 1 million, a sharp decline from its peak, during Herod's reign. The Jewish people had shattered into tiny enclaves throughout the Mediterranean basin and along trading routes in Europe, India, and the Silk Road into China. The number of Jews worldwide probably numbered no more than 2 million. While some Jews who fled rural Palestine established new lives in their new nations as farmers and cattlemen, others relocated to growing urban centers to become bakers, tradesmen, and merchants. As the importance of Palestine faded, the epicenter of Judaism split south, east, and west. The map of world Jewry began to look like a destination map for Air Diaspora—a web of tangled routes emanating from a few major hub cities with small but growing Jewish populations.

Alexandria in Egypt and the already stable community in Babylon flourished as cultural and religious centers. Rome emerged as the Western center of Judaism. After his conquest of Palestine, Titus shipped would-be Jewish slaves through the ports of Puglia (ancient Apulia) on the southeastern Adriatic coast. A trove of documents unearthed in 2003 in Venosa (ancient Venusia), a strategic gateway in the boot of the peninsula, indicates that five thousand

Jewish slaves were resettled there alone. "Claudia Aster, captive from Jerusalem," reads the inscription on one headstone. "Tiberius Claudius Proculus, imperial freedman, took care of this epitaph. I ask that you make sure through the law that you take care that no one casts down my inscription." It's reported that Jews did not make very good slaves because of their strict dietary laws and refusal to work on the Sabbath. Many owners were glad to be rid of them for a price, and the established Jewish community of Rome often obliged by buying their freedom.

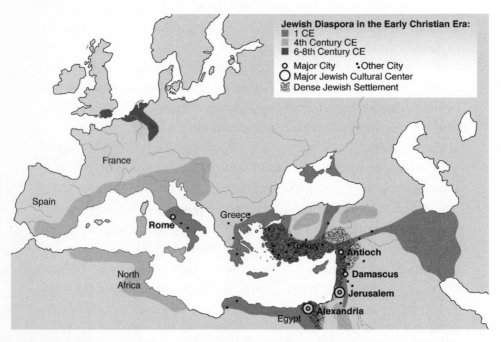

Figure 6.2. Jewish exilic communities after the destruction of the Second Temple.

Beleaguered Palestinian Jews found the Roman capital far more tolerant than the land they had been forced to flee. Jewish proselytizers even had some success in attracting converts from the Roman upper class. During Trajan's rule, from 98 to 117, three descendants of Herod became senators, the top echelon of Roman society. Flavius Josephus wrote his major works at the Roman imperial court. In 212, Emperor Caracalla issued his famous edict, allowing free Jews in the empire to become full Roman citizens, although as a consequence of the first Jewish Revolt, they were still required to pay a poll tax. Pious Jews were exempted from certain municipal obligations that involved pagan rites. While epitaphs from this period indicate there were Jewish painters, physicians, actors, and poets, for the most part, like most of the populace, Jews were poor, scraping by as butchers and shopkeepers. The first-century Roman poet Martial complained that it was difficult to sleep in the capital city because of the noise caused by Jewish peddlers and beggars. Jewish women were renowned as fortune-tellers. The number of destitute Jews far outnumbered the modest working class.

By the second and third centuries, Christianity was beginning to attract sizable numbers of converts from what some viewed as a sclerotic Judaism, riddled with cults and compromised by a divided rabbinical leadership. Christianity's appeal was in the transformative effect it had on people's lives, not in its doctrine, for there was none for centuries. There was no official New Testament. The early Christians fervently embraced a myriad of scriptures, very few of which would be recognized as Christian theology today, and many that advocated what most contemporary Christians would deem heretical. Because many early Christians were pagan converts, some Christian sects believed in multiple gods—two, twelve, thirty. Many viewed our temporal existence warily, embracing a devil-like-creator myth. Some sects maintained that Christ's death and his supposed resurrection had nothing to do with human salvation. Others questioned whether he really died—or lived at all.

The growth of Christianity was astounding. The tiny band of Jewish heretics had grown to one thousand by 40 CE. By the end of the first century, the new Christian sect had ten thousand followers, far fewer than the number of Jews and a tiny fraction of the estimated 60 million people in the vast Roman Empire. A century later, two hundred thousand people called themselves Christians. By 300, the number had exploded to more than 6 million. The attitude of the Romans toward Christianity inexorably evolved from persecution, to disdain, to tolerance, and ultimately to conversion.

From the time of Paul onward, Jewish rabbis increasingly found themselves on the defensive in rancorous public debates over the teachings of the prophets, whose ambiguous words many Jewish Christians took as predictions of Jesus's divinity. Isaiah, Jeremiah, and Ezekiel were considered apostles of the "true" Israel. The gospels gradually came to be seen not as spiritual metaphor but as literal history. It was the kind of certainty that attracted loyal followers among the spiritually unsettled put off by the disputatious character of Judaism, which constantly debated the twists of biblical metaphors.

The defining event in the diverging fortunes of Judaism and Christianity came in the early fourth century. Constantine, a thirty-two-year-old provincial Roman leader in the western half of the empire, defeated his regional rival Maxentius at the Milvian Bridge, near Rome, in 312. According to Christian tradition, the letters XP ("Chi-Rho," the first two letters of "Christ") in Greek intertwined with a cross appeared before him in the sky. Inspired by the vision, he defeated his formidable rival. Supposedly, in response to the miracle, Constantine issued what became known as the Edict of Milan. Christians in the western sections of the empire would no longer be tortured or killed. The emperor of the eastern provinces, which included Palestine, issued a similar proclamation but soon rescinded it. Constantine then executed a series of daring moves designed to politically and religiously unify the vast empire. He launched a military campaign that overran the east. To shore up support, he restored the property and power of Jerusalem's Christian bishops. In 324, he transferred the capital of the entire empire to New Rome, Constantinople, the site of the ancient Greek city of Byzantium.

At the time, Christianity was still split by doctrinal disputes. Although many, if not most, Christians believed that the resurrection literally happened, this was by no means universally embraced. Christian doctrine would be formalized as much by political decree as by religious revelation. To complete his grand plan of unification, Constantine convened a conference in ancient Nicea, now the

Turkish town of Iznik, to settle the matter. What came to be known as the Council of Nicea drew around 250 bishops, mostly from the eastern provinces. With Constantine presiding, Christianity was consecrated as the official religion of the Roman Empire.

Most fateful for Jews, the council dramatically inflated the significance of the crucifixion and reinforced Christian claims for Jesus as the Son of God. The new orthodoxy proved catastrophic for Jews. Although Judaism was not declared a "prohibited sect," fervent Jews came to be seen as backward and superstitious and were occasionally targeted for their perceived role in Jesus's killing. The Hebrew Bible was rechristened the "Old Testament." Subsequent councils forbade Christians from celebrating Passover, although many ignored the edict at first. The alternative Christian celebration of Jesus's resurrection was later named Easter after the Teutonic pagan goddess of the rising light of day and the spring. Christians were banned from observing the Jewish Sabbath, as Sunday became the Christian day of prayer.

Jews were also forbidden from possessing Christian slaves, an attempt to curtail the common practice of taking non-Jewish wives or concubines. The law, which prevented Judaism from competing for converts, ironically promoted Jewish genealogical "purity," because by the custom of the times, wives automatically became Jews, as did slaves when enfranchised. The enforced separation of Jews and Christians was formalized in 388, when the empire passed a law making intermarriage with Jews a capital offense. Many Jews, increasingly seen as social outcasts, assimilated or converted. From this point forward, Christianity and Judaism would evolve along distinctly different biological as well as religious and cultural tracks.

With Constantine diverting resources to Constantinople, Christianized and Greek-speaking Byzantium thrived, while the western empire fell into steady decline. Rome collapsed in the fifth century as barbarian Huns, Visigoths, Ostrogoths, Vandals, Franks, and Burgundians carved up the Italian provinces into minor fiefdoms. By the seventh and eighth centuries, financially weakened and challenged on its western and southern borders by the sudden rise of Islam, the Byzantine Empire would relinquish much of its territorial gains. The major cities except Constantinople would gradually fade away to small, fortified centers, and the centralized military would fracture into local armies.

With an actual homeland no longer the central motif of their religion, the Jews of early Christendom retreated ever further into scripture, the surviving rock of their religion. Unable to worship in the splendor of their Temple in Jerusalem, they were united not under the flag of nationhood but by a belief in the Torah. Judaism was crystallizing into a religion of history and fidelity defined as "in the blood" and "of the Book." The descendants of the ancient Israelites were now mostly strangers in strange lands. Exile and the image of the "wandering Jew" were by now firmly embedded in Western mythology.

CHAPTER 7

WANDERING TRIBES

I'm certain the Lost Tribes of Israel are alive and well. I know, for example, that the exiled tribe of Dan sojourned across Europe and named many rivers, towns, and countries during its travels, such as the Danube River, Donegal, and Denmark. The tribe of Ephraim settled in Britain; the word "British" is of course derived from the ancient Hebrew word *beriyth*, which means "covenant." The Israelites, "Isaac's sons," naturally became known as the Saxons. And Queen Elizabeth and her lineage are direct descendants of King David and the legitimate heirs to the Israelite throne. It's all clear.

How do I know this? Because the Worldwide Church of God tells me so, or at least it used to. That's the Pasadena, California, sect founded by an ex–advertising agent, Herbert W. Armstrong, in the early 1930s.

Armstrong was a devout believer in what is called British-Israelism or Anglo-Israelism, a movement originating centuries ago and still alive today that maintains that most Britons are descendants of ancient Israelites. Armstrong got his start in 1931, when he launched the hit program *The World Today* on the Radio Church of God. His message mixed elements of Judaism with a dash of Seventh-day Adventism and a sprinkling of pagan mysticism. He cooked up quite a stew. He preached that the Bible was "a coded message not allowed to be revealed"—that is, until he cracked its secrets. He was willing to share its hidden wisdom with anyone and everyone, particularly those who would cough up 25 percent or more of their income to help spread "the word." When Armstrong died in 1986, the WWCG claimed more than 150,000 members and an annual budget of $130 million. His successors ultimately abandoned Anglo-Israelism for an evangelical Protestantism, although numerous Armstrong family members and friends founded sects that preserve some of the zanier claims of the elder sage.

The search for the missing Lost Tribes ranks right up there in biblical mythology with the quest to find the Ark of the Covenant and the Holy Grail. The mystery of their whereabouts has encouraged a motley crew of true believers, mystics, zealots, troubadours, and out-and-out fakers. It is so alluring and central to questions of Western identity that an equally unusual assortment of truth-seekers has more recently joined in the quest: anthropologists and geneticists.

According to the Hebrew Bible, Jacob, later renamed Israel, was the father of the original Twelve Tribes of Israel. The distinctive emblems of each tribe were carried on the vestments of the Cohen Gadol (the high priest), the Levite descendants of Aaron. Other than Jacob's words in Genesis 49, there are scant references to the personalities and responsibilities of these tribes in the Bible, so speculation and myth have filled in the gaps.

It's assumed, based on the practices of the time, that the tribes kept to their circumscribed regions and didn't intermarry much. Each tribe had a reputation for special expertise. Many of the national leaders supposedly came from the tribes of Judah and Benjamin in the south and Ephraim in the north; Benjaminites were great hunters with a fighting streak; Zebulunites, located on the seashore, traveled the world as traders; Menasseh raised cattle and were often warriors; descendants of Reuben, Jacob's firstborn, were known for their fiery temperaments and apocryphally were the seeds of the French; the Issachar, whose ancestor Job, whom the ancient Greeks called Cheops, supposedly had a hand in the construction of the Great Pyramid, were said to be destined to be either slaves or engineers; members of Dan were characterized as serpents, which by tradition meant they were quick to seek judgment in court and were sometimes said to be the source of today's stereotype of the "Jewish lawyer"; the sons of Asher, which means "rich" in Hebrew, supposedly produced oil and other products of royalty; Simeonites were warriors, which may have led to their extinction; the Ephraim were known as colonists, a designation likely conferred upon them post hoc for allegedly settling in Britain; Naphtalites were known for their spirituality and elegance as well as lack of practicality; the Gad were raiders and uncompromising in their religiosity; and the Levite subgroup of Cohanim were known for their spiritualism, as they were responsible for the Temple service and religious instruction.

In Genesis, Jacob bids his sons farewell before he dies, prophesying their precarious futures as tribal leaders. Their fate has been the focus of the greatest search effort in human history. Unity eluded the children of Israel. The first wave of Jews left when the tribes fractured into northern and southern kingdoms after the civil war in the time of Solomon's son, Rehoboam. The Bible subsequently tells of numerous instances when sizable numbers of Hebrews were expelled or forced to flee from their homeland. "Israel are scattered sheep, harried by lions," says the prophet Jeremiah in Jeremiah 50:17. "First the king of Assyria devoured them, and in the end King Nebuchadnezzar of Babylon crushed their bones."

While it's well documented that many Judeans in the south ended up in Babylon after Jerusalem fell in 586 BCE, an enduring mystery persists over the fate of the Israelites of the northern kingdom. The exiled tribes wander off the pages of the Bible and leave no trace in the historical record. The next extant reference to them appears more than five hundred years later in Josephus's *Antiquities of the Jews*, in which he casts the fate of the northern Israelites in mythic but vague terms. "There are but two tribes [Judah and Benjamin] in Asia and Europe subject to the Romans," he wrote, "while the

Ten Tribes are beyond the Euphrates till now, and are an immense multitude, and not to be estimated by numbers."

Did God promise that the descendants of this "immense multitude" would be restored to biblical *terra sacra*? "He will hold up a signal to the nations and assemble the banished of Israel, and gather the dispersed of Judea from the four corners of the earth," says the prophet Isaiah in 11:11–12. Ezekiel expresses a similar sentiment in 34:12–13: "As a shepherd seeks out his flock when some in his flock have gotten separated, so I will seek out My flock, I will rescue them from all the places to which they were scattered on a day of cloud and gloom. I will take them out from the peoples and gather them from the countries, and I will bring them to their own land, and will pasture them on the mountains of Israel . . ."

While the prophets certainly believed that redemption was contingent on the return of the lost Jews, not every Jew and Christian today agrees—nor did they two thousand years ago. As might be expected considering the traditional Jewish love of a good debate, the Jews of the early Roman Empire were divided over whether to take the predicted return on its face. A famous midrash in the Talmud portrays a fierce dispute over this very issue between two early-second-century rabbis. According to Rabbi Eliezer, their return was indisputable: "Like as the day grows dark, and then grows light, so also after darkness is fallen upon the Ten Tribes, God will ultimately shine light upon them."

He was challenged by the renowned Rabbi Akiva (who laid the foundation for the Oral Law, codified later as the Mishnah, and was the spiritual leader of the Bar Kochba revolt), who was worldlier as the result of his extensive travels throughout the Roman world. "The Ten Tribes shall not return again," he wrote, "for it is written: 'And he cast them into another land like this day.' Like as this day goes and returns not, so do they go and return not." When Rabbi Akiva had happened upon Jews during his treks through Armenia, Kurdistan, and across the Mediterranean basin, he saw these communities for what they likely were—not biblical exiles, but Jewish traders who had put down roots in distant lands.

For the Jewish people living within diaspora communities that numbered in the hundreds, the Lost Tribe stories provided solace during times of persecution—the promise of a return to a future homeland and ultimately salvation. But it is in Christianity that the fascination with the biblical exile has taken deepest root. Today, wherever you find a tribe or ethnic group proclaiming their Israelite heritage, you are likely to find the fingerprints of Christian missionaries or biblical romantics. Some Christians still dutifully embrace Jesus's words to his twelve disciples in Matthew 10:5–6: "Go nowhere among the Gentiles, and enter no town of the Samaritans, but go rather to the lost sheep of the house of Israel." Jesus expected that command to be honored in death as well. According to Matthew 19:28 and Luke 22:30, in the afterlife, Jesus's disciples would preside over the Twelve Tribes. Some scholars see these passages as echoes of the dream sequence in Daniel 7, in which a humanlike being identified as the "Ancient of Days"—which some Christians interpret as being Jesus—materializes from the clouds to take dominion over the four kingdoms of the earth.

Is this gathering really supposed to occur? Biblical scholars believe that Revelations, while ascribed to John, may have been written by a Palestinian Jewish Christian who fled into the diaspora as a consequence of the first Great Revolt against the Romans (66 to 73 CE), suggesting that the gathering together of the tribes will occur in heaven, not on earth. This "new Jerusalem" has "twelve gates" on which are "inscribed the names of the twelve tribes of the Israelites," reads Revelation 21:12, 21. "And the twelve gates are twelve pearls, each of the gates is a single pearl, and the street of the city is pure gold, transparent as glass." Although these passages hint that these tribes are more symbolic than literal, over the centuries, many Christians, like many Jews, have expected to find remnants of these tribes.

The Christian belief in the return of Jesus is inextricably tied to the historical Christian mission to rescue missing Jews, return them to Israel, and convert them. The image of the lost Jew, his soul in purgatory, has been emblazoned in Christian consciousness.

According to a popular and pernicious French medieval embellishment of this stereotype, Jews were condemned to wander the earth until the Second Coming because a Jew had supposedly refused to allow Jesus to rest his cross at his door on the way to Calvary. Christian fundamentalists have been convinced that the discovery of the missing Israelites will affirm the coming of the End of Days and the promised redemption. Their millennialism has rested in part on their belief, derived from Paul's Epistle to the Romans, chapter 11, that God will convert the Jews to Christianity as a prelude to the Great Kingdom, although many Jews will apparently ignore the messianic call. In Romans 9:27–28, Paul reiterates Isaiah's prophecy that "only a remnant" of the exiled Israelites will be saved by embracing Jesus as their savior. Even in church services today, Presbyterians are officially commanded to pray for the conversion of these lost souls.

The missing Jews fell into three categories, much as outlined in the Talmud, which predicted a full accounting of the tribes when the Messiah returns: pocket communities still living as Jews but in contact with major Jewish urban centers; the crypto-Jews who had assimilated into gentile society but retained some Jewish practices that were often a mixture of Christian or pagan rites; and those Jews who were truly lost. Across centuries and continents, prophetic words instilled hope that one day these scattered communities, defined by a belief in One God's commandments, would be reunited in the Promised Land. Today, hundreds of religious sects with no direct link to modern Judaism claim ties to the ancient Israelites, offering romantic tales of their origins. This presents a fascinating dilemma. To what extent are the predictions of the holy texts allegory or history, a product of faith or rather the historically faithful?

THE SEARCH COMMENCES

The Bible provides only a murky account of where these lost Jews may have relocated. The Lord removes the Israelites from

His sight—which we can assume to mean far away and without communication with the motherland. First Chronicles 5:26 and 2 Kings 17:6 and 18:11 hold that the victorious Assyrian king Tiglath-pileser dispatches the tribes of Reuben and Gad and half of the tribe of Menasseh from the northern kingdom of Israel to Halah, Harah, Habor, the cities of the Medes, and to Gozan. Where are those places? If these places of exile really existed, they were probably once towns in the vast Assyrian Empire. Many biblical scholars looking for the homes-in-exile of these displaced tribes and all of the rest of the ten tribes of Israel subsequently deported by Sargon upon the fall of Samaria have pointed to Habor as the "Chebar Canal" mentioned in Ezekiel 1:3. They believe it may refer to the confluence of the Euphrates and Tigris in modern-day Baghdad, which became home to one of the world's oldest continuously occupied Jewish enclaves. The origins of that community are unclear, however. When the Judeans were exiled to Babylon 138 years later, they apparently did not find any Israelites in the area.

Because none of the prophets was particularly helpful in identifying actual places where these exiled Israelites lived, a Lost Tribe search industry, if you will, materialized. The first renowned seeker was Eldad Ha-dani, a mysterious Jew who made his way through the courts of North Africa and Iberia in the late ninth century, regaling all who would listen with stories of wandering Jews. Under siege from the rising tide of Islam, the Christian world at the time was awaiting Jesus, while Jews pined for a messiah of their own. Eldad preached to the desperate. He turned up first in Tunisia in 883, speaking Hebrew and declaring that he was descended from the tribe of Dan, which he said was flourishing in a land known as Cush (identified by biblical historians as Ethiopia or Sudan). According to "the Danite" (as he was also called), his tribe had left Israel well before the Assyrian and Babylonian expulsions, when the United Monarchy first fractured after King Solomon's death. Three other tribes—Naphtali, Gad, and Asher—soon followed the Dan. After Nebuchadnezzar leveled the First Temple in Jerusalem in 586 BCE, the four exiled tribes were joined by the Levites.

How did these "Sons of Moses" supposedly get to Africa? God had transported them, magically. Together, the tribes supposedly lived an Edenic life on the banks of an amazing river, the Sambatyon, named after the Sabbath because it stops flowing for one day each week. This myth of the Sambatyon River had been knocking around for hundreds of years. In differing accounts, the river flows with foaming water or rocks that make a terrible rumbling sound heard for miles around. The Roman historians Pliny the Elder and Flavius Josephus had each referred to it in their first-century writings as the Sabbatic River. The tribes remained insulated from invaders, legend holds, because the river would periodically turn into a fiery barrier on the one day each week that it stopped flowing, shrouding them in a thick cloud.

Eldad was given a lavish welcome wherever he went and reportedly made a fortune from his courtly connections. Then he disappeared as mysteriously as he had appeared. But his legacy lived on. In later embellishments of his tall tales, the tribes of Issachar, Zebulun, and Reuben were located near the mountains of Paran, most probably the Caucasus Mountains; Ephraim and one half of the tribe of Menasseh turned up in southern Arabia; and the other half of Menasseh and the tribe of Simeon were supposedly discovered in the "land of the Khazars," the plains north of the Caucasus. Eldad's stories, however fantastic, provided a new atlas of the diaspora, and his notoriety offered inspiration, however implausible, for Lost Tribe seekers for years to come.

Three centuries later, two fabulists, one Jewish and one Christian, would help burn the legend into the popular imagination. If his diaries can be believed, beginning around 1165, Benjamin of Tudela, a Jew from the Basque region of Spain, embarked on a thirteen-year trek through the Jewish world, from Europe, to Cyprus, to Baghdad, to the western reaches of the Chinese empire. Benjamin reported finding missing Jews everywhere. Some of his accounts appear reasonable: 3,000 Jews in Constantinople and 200 in Jerusalem. Others seem less so: 300,000 Jews in Arabia, including 50,000 in Yemen, who were supposedly remnants of Reuben, Gad, and half the tribe of Menasseh. He also located the tribes of Zebulun, Dan, Asher, and Naphtali in Persia, where the Assyrians supposedly had relocated them.

Were his claims of finding Jews fabricated? Maybe not. Inscriptions found in the fifth and sixth centuries support the existence of a Jewish kingdom in Yemen's Samin Mountains and another in Ethiopia that battled Christian and pagan tribes and forced some of them to convert to Judaism. More than likely, his stories were a hybrid of hazy historical events and transmogrified versions of Eldad's tales.

About this time, the first popularized Christian version of the Lost Tribe tales began to appear throughout Europe. According to legend (and depending on who was telling the tale) a certain Presbyter Johannes, Priestly Juan, or Prester John ruled over a vast and wealthy empire as (choose one) king, pope, saint, caliph, or priest in (select one) India, Ethiopia, Asia, or Africa after converting his people to Christianity during the tenth, eleventh, or early twelfth century. More than one hundred manuscripts recounting versions of the Prester John story are preserved in libraries around the world in Hebrew, Latin, Arabic, Spanish, German, and other languages.

The story is probably a conflation of both Christian and Jewish legend. Early Christian lore had the original Apostles marching around the world from Ethiopia to India to China to spread the gospel. It's not hard to imagine how this improbable tale of theological colonization by Jesus's loyal followers blended with the Jewish story of the Tribes. The Prester John texts all reference the lost Israelites, the myth of the Sambatyon River, and a place called Paradise. The Tribes, along with Amazons and Pygmies, flourished in a land not unlike Eldorado or Lake Wobegon, with friendly animals, rivers that overflowed with precious stones and gold, and "beautiful women . . . ardent by nature." But watch out for those crafty Jews, Prester John warned: "We are placing guards at the passages, for if the Jews were able to cross they would cause great damage in the whole world against Christians as well as Ishmaelites and against every nation and tongue under the heavens, for there is no nation or tongue which can stand up to them."

The Prester John letters, which appear to borrow heavily from the Eldad fictions, had a powerful effect on the Christian imagination. They served dual and often-contradictory purposes of representing a fictionalized warning about the vaguely threatening Jews of Christian Europe and the incarnation of hope for a better future represented in the Second Coming.

The tall tales reappear in slightly altered form in the fourteenth century. The Venetian traveler Marco Polo, who spent a quarter century on the Silk Road in Asia, found Jews in "Middle India" with a mysterious mark on their cheeks and a "David King" ruling over "Georgiana." Around the same time, a writer claiming the nom de plume of Sir John Mandeville penned a wildly popular, half-baked compilation known as *Mandeville's Travels*. The mysterious stories told of "ten lineages" of Jews in the mountain valleys of Scythia in a distant land beyond Cathay, an ancient name for China. This Mandeville character clearly did not think much of Jews. He translated Prester John's fear of "the other" into a more demonic vision, claiming that a Jew had confessed to him that all Christendom was being targeted for poisonings. This libel quickly made the rounds in a Europe beset by the Black Death.

Christian credulousness over the story of the Lost Tribes took a bizarre turn in the early sixteenth century, when a swarthy, waif-like Jew in Oriental dress from Arabia calling himself David the Reubenite (also known as Reubeni) turned up in Venice around 1524, claiming to be a prince of the Israelites. He entranced local Christians with the fantastic tale that he was sent by his brother Joseph, the king of the tribe of Reuben, to obtain arms and money to fight off the Muslim Turks. Somehow, the story was believed, and he was sent off to Rome to see the pope. Upon arriving on a white horse, Reubeni was given a warm reception by Clement VII, whom he serenaded with stories of the descendants of Abraham cavorting on the banks of the Sambatyon. He timed the con well, for Christendom was under assault by the Ottoman "hordes." He convinced the pope and some Judeophile cardinals that Christianity desperately needed Joseph's army, which Reubeni claimed topped 300,000 Arabian warriors. Intrigued, Clement arranged for him to visit the king of Portugal, whom the pope thought had the means to best facilitate the alliance.

Arriving on a ship flying the Star of David, Reubeni visited King João III, who received him with great pageantry. Reubeni claimed that Joseph reigned over two and a half tribes, while the remaining lost Israelites lived in Ethiopia. The king promised him eight ships and supplies, but then the plan imploded. Portugal was home to tens of thousands of *Nuevo Cristanos*—New Christians—Jews, mostly from Spain, who had abandoned their faith, willingly or under threat of death, during the Inquisition. Many of these *conversos*, who had continued to secretly practice Judaism, went into a messianic frenzy when word leaked of a mystery Jew received by the king with such pomp. Fearing chaos, the king summarily ordered Reubeni out of the country.

Reubeni's quixotic cause took a fateful turn when he hooked up with a Portuguese dreamer and cabalist named Diego Pires, a New Christian who had also been thrown out of Lisbon, accused of stirring up other recent converts. A trek through Jerusalem had convinced Pires that he was the Messiah, whereupon he began preaching full-time under the name Solomon Molko. After being turned away by papal authorities in Bologna and Milan, the bizarre duo made a deadly mistake when they traveled to Ratisbon (now Regensburg in Germany) to present themselves to Emperor Charles V. Charles reportedly thought both of them were deranged, which they probably were. Molko was burned at the stake in 1532, while Reubeni was dispatched to a prison of the Holy Inquisition in Spain, where he died in 1535.

THE BRITISH ROMANCE WITH
JUDAISM

In 1559, Queen Elizabeth decreed what became known as the Religious Settlement, which restored the supremacy of the Protestant Church in England. That upset many Presbyterians, Independents, and Baptists, who were inclined to find comfort in the tall tales of the Lost Tribes adventurists. Although there had been no Jewish communities in England since a formal ban in 1290, the fast-growing Calvinist Protestant movement viewed them as the key to salvation and the promised era of peace. Some early Puritan leaders circulated petitions advocating asylum for Jews being persecuted in Europe, a proposal not entirely altruistic, as rescued Jews would be easier to proselytize and convert. They cited Romans 11:25, which they believed foretold of a mass conversion of Jews to Christianity just before the End of Days. The next edition of the Geneva Bible, issued in 1599, would categorically embrace this Puritan reinterpretation of the text.

It wasn't long before Christian Europe was in full Lost Tribe fever. A burning desire to convert lost souls led to Lost Tribe sightings in any number of places. Giles Fletcher, Queen Elizabeth's envoy to Moscow, was certain that the mountain region near the Caspian Sea, home of the extended clan of Genghis Khan, was the Medes mentioned in the Bible. Fletcher assumed that the Tatars were the returning kings mentioned in Revelation 16. There were any number of explanations for his sightings: the Jews could have been remnants of the long-defunct Jewish kingdom of Khazaria; so-called Mountain Jews of the Caucasus, who have claimed that they are descended from Judeans deported to central Asia by Nebuchadnezzar; Tadjiki Jews of Bukhara, who today refer to themselves as "Isroel" or "Yehudi" (and consider the Uzbek capital of Samarkand to be a twist on the name "Samaria"); or the Ebraeli Jews of the mountainous countryside along the Black Sea in Georgia, who claim links to both the Assyrian and Babylonian exiles.

Whatever Christian England thought of the Jews in the seventeenth century—and Jews were both romanticized and vilified, sometimes simultaneously, such as in Shakespeare's *Merchant of Venice*—everyone agreed the return of the Lost Tribes could hasten the Second Coming. Many in England came to believe they were descendants of the tribe of Ephraim, blood descendants of the Jews and inheritors of the covenant of Abraham. Some Puritans even thought they were reincarnated ancient Hebrews and chose biblical names for their children. A large faction embraced millennialism and became Christian Zionists. In 1621, the attorney Henry Finch, interpreting Genesis 12:3, argued that God would bless those nations that supported the rebirth of a Jewish Israel. "They shall repair to their own country, shall inherit all of the land as before, shall live in safety, and shall continue in it forever," he said in a parliamentary address.

The Scottish clergyman John Drury spoke for many churchmen in the early years of the Protestant Reformation when he argued that for the End of Days scenario to play out, there would have to be Jews in every corner of the world, which meant they should be readmitted to England. Drury drew on the work of Menasseh ben Israel, a prolific seventeenth-century Hebrew scholar of Spanish ancestry whose family had returned to Judaism after a forced conversion during the Spanish Inquisition. Living and writing in Holland, he argued that the tribes existed in England's colonial backyard, the Americas.

The Lost Tribe excitement spread to the Puritan-led government under Oliver Cromwell. In 1653, Cromwell cited God's message to Israel in the sixty-eighth psalm to claim that he and his supporters were chosen by God to preside over the establishment of His rule on earth. Cromwell's rump "Little Parliament" briefly replaced the English constitution with the Hebrew Bible's Laws of Moses. In 1655, after fierce lobbying by Menasseh, Parliament voted to rescind the ban on Jews. The vote stirred violent passions, with charges that Cromwell was secretly Jewish and that his brethren, on their return, would buy Saint Paul's Cathedral, but the law was passed, thanks to the power of the Lost Tribes legend.

British Israelism resurfaced at the end of the eighteenth century in the wake of the American and French revolutions and as Britain embarked on colonial expansion. Lord Shaftesbury, a conservative evangelical Christian and an intimate of leading members of Parliament during the 1830s and 1840s, emerged as an influential Christian Zionist and is often cited as an inspiration to Jewish thinkers, including Theodor Herzl. In 1839, he coined the phrase "a people with no country for a country with no people," arguing in a famous article that "the Jews must be encouraged to return [to Palestine] in yet greater numbers and become once more the husbandmen of Judea and Galilee."

John Wilson is viewed as the modern father of the movement. In *Our Israelitish Origins*, published in 1840, he provided a "scientific" foundation for the British obsession, theorizing that modern Europeans, particularly Anglo-Saxons, were descended from certain Scythian tribes, who were in turn descended from the Lost Tribes. Wilson's preaching didn't get much attention until his message changed from claiming an Israelite heritage to the more useful notion that Britain possessed the Abrahamic Covenant and was itself the new Promised Land. Edward Hine cited Wilson as his inspiration for his 1874 book *Forty-Seven Identifications of the British Nation with Lost Israel*, in which he correlated biblical Israel with the British Empire of his day, citing vague references in Jeremiah to the "isles far off" and the "north country." Hine later toured the United States on a mission, not entirely unsuccessful, to convince the native Indian population they were descendants of one of the Lost Tribes, Menasseh.

British Israelism reached its apogee in the mid-twentieth century. In the introduction to his 1978 book *The Genetics of the Jews,* Arthur Mourant wrote movingly of his youthful romanticized notion of the "Jewish race." Although raised as a fundamentalist Methodist, Mourant became obsessed with Israel as a young boy. He recalled with fondness his first schoolteacher, a "British Israelite who believed that all Jewish people, if not all Europeans, were descended from the 'Lost Tribes of Israel' . . . [F]or some of my most formative years," he wrote, "I considered myself a Jew, and this sense of identity has persisted [even] though I can now see no evidence that I have any Jewish ancestors." Years later, the influx of Jews to Palestine and later to the State of Israel appeared to fulfill the biblical prophecy of the book of Revelations.

Such views have never gone completely out of fashion in Britain, where some still consider the monarchy an extension of the Davidic Empire. The echoes of Anglo-Israelism can be heard even today at many events marked by the singing of William Blake's poem *Jerusalem*, including at the annual conference of the British Labour Party. The glorious nation rising in the fertile field may seem like a parable of England's greatness, but for Blake, it was a statement of his religious conviction. As a dedicated Anglo-Israelite, he viewed Britain as New Jerusalem, where the British, the descendants of the Hebrews, would gather at the Second Coming.

COMING TO AMERICA

Promised Land imagery was deeply embedded in the American imagination. The Pilgrims fancied themselves as God's chosen people and saw America as New Canaan, a new home for a lost tribe. "Come, let us declare the word of God in Zion," declared William Bradford upon landing on Plymouth Rock. They named their children after key figures in the first five books of the Bible and called more than one thousand of their towns after biblical places, including Bethel, Bethlehem, and, of course, New Canaan. "We shall be as a City upon a Hill, the eyes of all people are upon us," wrote John Winthrop of the Puritans in 1630.

To justify expelling the Native Americans from their land, the English pioneers fancied themselves New Israelites confronting the pagan Canaanites who had to be destroyed for destiny to be fulfilled. In contrast, and almost as powerful, was the myth promoted by some Christian missionaries that Native Americans were remnants of the Lost Tribes. Bishop Bartolomé de las Casas, a contemporary of Columbus's known as the "Apostle to the Indies," believed that the Native Indians of Santo Domingo spoke a "corrupt Hebrew." Spanish explorers turned up more "lost Jews" in the Yucatán (where the natives abstained from pork and embraced circumcision) and Peru (where the inhabitants supposedly had large noses, spoke through their throats, and rejected Jesus).

A slew of treatises and books appeared in the eighteenth and nineteenth centuries, sparked by the writings of the pioneer Indian trader James Adair, who noted twenty-three dubious parallels between Indian and Jewish customs, for instance, claiming the Indians spoke a corrupted form of Hebrew and offered animal sacrifices. William Penn, an English Quaker, saw signs of the ancient Israelites in the Lenape Indians. "I am ready to believe them of the Jewish race, I mean of the stock of the Ten Tribes," he wrote in 1683, citing a list of rituals with supposedly Jewish origins: "They agree in rites; they reckon by moons; they offer their first fruits; [and] they have a kind of Feast of Tabernacles." There were so many Jews around, he said, that it was like being in the Jewish Quarter in London.

The belief in an Indian-Jewish connection has endured. Mel Brooks fans no doubt fondly recall the 1974 western spoof *Blazing Saddles*, in which the director made a cameo appearance as a Sioux Indian. The scene looks at first as if it's from a classic western. A hapless family in a covered wagon has lost its way in the desert, baking under a noonday sun. All of a sudden, an Indian war party descends on the weary travelers. An elaborately decorated chief—Brooks—comes to confront the haggard and terrified stragglers. With the tension building, the chief starts speaking in his native tongue—Yiddish! It was classic Brooks shtick, but to those with knowledge of the quest for the Lost Tribes, it was a delicious double entendre firmly rooted in history.

These contradictory sentiments—both the immigrants and the Indians were said to hold claim to the Israelite tradition—helped fuel the myth of American destiny. In 1776, Benjamin Franklin and

Thomas Jefferson wanted Promised Land images for the new nation's seal. Franklin proposed representing Moses dividing the Sea of Reeds, only to swallow up pharaoh's army, while Jefferson argued for portraying the Israelites being guided in the wilderness by the pillar of fire by night and the clouds by day. "I shall need . . . the favor of that Being in whose hands we are, who led our fathers, as Israel of old, from their native land and planted them in a country flowing with all the necessities and comforts of life," Jefferson would declare in 1805 in his second inaugural address.

Images of the Promised Land remain central to the Church of Jesus Christ of Latter-day Saints (LDS), the Christian-like religion founded in 1820 by Joseph Smith, who had been a Methodist farmer. Smith claimed that God had carved the religion's seminal text, The Book of Mormon, on gold tablets, which he found at God's direction but which then mysteriously disappeared. According to LDS theology, after the Tower of Babel was destroyed, the brother of Jared led his people to the land "where never had man been," which Smith maintained was the Americas. Around 600 BCE—shortly before the Assyrians began dismembering the northern kingdom—a wealthy Hebrew merchant from the northern tribe of Menasseh named Lehi fled with his friend Ishmael and their families south, into the Arabian Peninsula, until they reached a fertile coastal region called Bountiful. They soon built a ship and sailed across the eastern oceans to the Americas, where Lehi's sons, Nephi and Laman, were said to have founded rival tribes: the Nephites and the Lamanites. After centuries of peaceful coexistence, the Lamanites turned on their sibling nation, all but wiping them out.

In the Book of Mormon, the Lamanites are described as "cursed" by God as punishment for their wickedness and corruption: "And he had caused the cursing to come upon them, yea, even a sore cursing, because of their iniquity," reads 2 Nephi 5:21. "[W]herefore, as they were white, and exceedingly fair and delightsome, that they might not be enticing unto my people the Lord God did cause a skin of blackness to come upon them." While most Mormons consider this passage to mean that Lamanites were the ancestors of

Native Americans, some members of the church hypothesize that Lamanites may have intermarried with indigenous Native American peoples with darker skin. Second Nephi 5:24 says that these dark-skinned descendants of the Hebrews were fated to "become an idle people, full of mischief and subtlety, and did seek in the wilderness for beasts of prey"—a familiar racist stereotype about Native Americans.

LDS missionaries still actively try to convert Native Americans so that they might be restored to their "original state." Church anthropologists have studied the monuments of the Aztecs and the Maya in the hope—not yet fulfilled—of establishing a historical link to the ancient Hebrews. The LDS Church also has built a research center in Jerusalem to search for connections they might have with peoples of the ancient world. In the LDS version of the End of Days, church members will be joined by the Lost Tribes, who Joseph Smith proclaimed in his divinely inspired *Doctrine and Covenants* would return to the New Jerusalem to be built in Jackson County, Missouri. For the time being, until Jesus returns to Zion, the remnants of the Ten Tribes remain holed up somewhere in the "north country," referred to by some LDS elders as the "frozen north" and the land of "Sinim," which is one of the domiciles-in-exile of the dispersed Israelites mentioned in Isaiah 49:12.

Echoing the claims of Judaizers around the world, representatives of the "Northern Cherokee tribes of the Old Louisiana Purchase" assert that they are the Israelites referred to in Mormon scripture. The Missouri Cherokees, as this nonsanctioned tribe calls itself, claim they are descendants of Jewish Zealots who the historian Josephus wrote fled Palestine after the failed revolt at Masada. The evidence? Jews at that time and the Cherokees of the Old Louisiana Purchase both braided their hair. "The story has been kept alive among our Cherokee people that the Sicarii [the Jewish partisans of Masada who concealed *sicae*, or small daggers, under their cloaks] . . . are some of our ancestors who managed to cross the water to this land, and later became known as Cherokees," a spokesperson for the tribe said, noting the phonetic resemblance of "Sicarii" and "Cherokee," which she pronounces as "Tsára-gí." Most historians believe the Missouri Cherokees are falsely claiming tribal status with hopes of somehow exploiting their supposed Jewish and Mormon connections.

Geneticists have now weighed in on the controversy over whether Native Indians are indeed descendants of ancient Semites. DNA samples taken from tribes in South, Central, and North America (though not from the Missouri Cherokees) have shown that their principal ancestors were from northeast Asia, not the Middle East. LDS authorities generally ignored the findings until 2002, when the Mormon anthropologist Thomas Murphy set off a firestorm in his article "Lamanite Genesis, Genealogy, and Genetics." The belief that American Indians came from Israel is tantamount "to claiming the earth is flat," Murphy concluded. "Many people would like to see the LDS Church publicly acknowledge that it is no longer appropriate to label Native Americans as Lamanites or heathen Israelites."

Defenders of LDS doctrine claim that Lehi's descendants could have been a small group whose genetic signatures were erased over time by genetic drift as they mixed with other indigenous American populations. Local church authorities initiated disciplinary and excommunication hearings against Murphy but backed off in the face of a huge outcry, including from some LDS scientists. The former Mormon bishop Simon Southerton, a devout believer but also an Australian molecular biologist, quit the church over the dispute and after his research found no evidence that "supported migration of Jewish people before Columbus." "The truth," he wrote, is that "there is no reliable scientific evidence supporting migrations from the Middle East to the New World."

AFRICAN JEWS

"Two broad categories of Jews [have] played a hugely important role in the Christian imagination," notes Tudor Parfitt, director of

the Center for Jewish Studies at the School of Oriental and African
Studies at the University of London. "There were the actual Jews
who lived, usually precariously, in European cities and who were
by and large despised and hated, and there were the *imagined* Jews,
of whom the Lost Tribes form a part and who usually provoked
more admiration and interest."

Parfitt speaks from experience. A prolific writer on the Lost
Tribes, he stands tall in the long historical tradition of legendary
seekers. "Western Christians, perhaps most acutely British and
American evangelicals, were prone to see traces of the Bible everywhere," he has written. But in key ways, Parfitt stands apart from
his predecessors. While many of them were sensationalist, Parfitt
is savvy and sober. And while almost all were driven by glory or
a messianic dream of salvation, he is motivated by a profound respect for the aspirations and identities of other cultures. But like
these legendary Lost Tribe seekers, Parfitt has landed in the same
end place as a true believer: he is convinced he has discovered in the
hinterlands of Africa a fragile twig of the Israelite genealogical tree,
surviving ever so precariously.

Parfitt's remarkable story began in 1984, when he was commissioned to write a report on the plight of the Beta Israel of Ethiopia,
popularly known as the Falasha. At the time, devastating seasonal
droughts combined with political ineptitude had brought millions
of Ethiopians to the brink of starvation. Among them was a group
of twelve thousand or more so-called Black Jews streaming over the
border into Sudan, claiming to be the lost tribe of Dan and desperate to make their way to their biblical homeland, although they had
no clue where to head. Their plight stirred an international rescue
effort, with many Jewish organizations pouring millions of dollars
into a massive airlift dubbed Operation Moses that transported,
almost overnight, a mythical Lost Tribe to Israel.

Much like non-Jewish Ethiopians, the Falasha believe they have
a biblical royal pedigree tied to the story of King Solomon and the
queen of Sheba, but with a crucial distinction. According to the
Ethiopian national epic, *Kebra Negast*, the queen returns from her

tryst in Jerusalem to ancient Ethiopia, where she bears their son, the future King Menelik. When Menelik grows up, he returns to Jerusalem and smuggles the Ark of the Covenant back to Ethiopia. With the ark in their control, the Ethiopians of mixed ancestry are transformed into the new chosen people. To underscore the supposed purity of their Israelite ancestry, the Falasha's origin story has them as descended not from King Menelik but rather from the firstborn sons of Israelite notables, members of the tribe of Dan sent by Solomon as a guard of honor to accompany Menelik back to Ethiopia. In 1975, Israel's chief rabbis, embracing the beguiling account and responding to the fervor of many Orthodox Jews, granted the Falasha official status as descendants of Dan. "I also wanted to believe the Falasha were a remnant of the ancient people of Israel struggling against all odds to rejoin their long-lost brethren," Parfitt told me.

History has provided only ambiguous clues. Jews were living in Ethiopia as far back as the second century BCE. The independent Ethiopian first century CE kingdom known as Aksum was more than likely suffused with Semitic or Judaic influences. The Falasha may have been a rump group that remained true to its historical roots when the Ethiopian king converted to Christianity in the fifth century. For centuries, the Black Jews maintained separate traditions from their Christian countrymen. While most Ethiopians ate raw meat, drank heavily, and rarely washed, the Falasha cooked their meat and were scrupulously sober and relentlessly hygienic. But their practices diverged from those of other diaspora Jews. Their brand of Judaism drew on a literal interpretation of the Hebrew Bible and showed no marks of the Oral Laws of rabbinical Judaism. Their central tenet was a return to Zion, which explains their desperate attempts over many centuries to immigrate to Jerusalem, the most recent during the famine crisis.

By the early 1990s, geneticists were finally in a position to test the Falasha's oral history against the hard facts of DNA. Alas, numerous studies have found no evidence that the Falasha are descendants of an Israelite tribe. The most definitive analysis, published in 1999 by Gérard Lucotte and Pierre Smets of the International Institute of Anthropology in Paris, found none of the most common Jewish genetic markers. "[T]he Falasha people descended from ancient inhabitants of Ethiopia who converted to Judaism," they concluded.

There are any number of small African tribes with similar jumbled origin myths, the best known being the yarmulke-wearing, Passover matzoh–baking Abayudaya of Uganda; the Nigerian Igbo (whose name is supposedly a corruption of "Hebrew"), who claim descent variously from the tribes of Ephraim, Menasseh, Levi, Zebulun, and Gad, practice circumcision, are loath to submit to the authority of a ruling class, and are known for their business acumen and technical prowess; the Mohammedan Berbers of West Africa, who are believed to be mixed descendants of Arabs or Jews expelled from Spain who then blended with the local population; the West African Fulani, a pastoral people with noticeably non-Negroid features, who believe they are descendants of an ancient Israelite herder who was expelled from a settlement; and the Lemba of South Africa and Zimbabwe, an obscure tribe known for its unusual religious practices. Most scholars were (and many still are) convinced that all of these so-called Black Jews—outcast tribes for one reason or another—adopted Jewish-like rituals when exposed to zealous Protestant missionaries intent on finding missing Israelites to convert to Christianity to hasten the return of the Messiah.

Parfitt had no reason to think differently until a curious string of events launched him on a remarkable journey of discovery. "I must say I didn't really believe this, this whole Jewish thing," he told me when we talked at his London office. The anonymous, tiny cubicle belied Parfitt's larger-than-life reputation as a modern-day Indiana Jones. He is a tall and sandy-haired Welshman, rugged and square-jawed, with an unapologetic fondness for jeans and safari shirts—the antithesis of an ivory-tower scholar.

Shortly after the publication of his book on the Falasha, *Operation Moses*, he was invited to speak at the University of the Witwatersrand in Johannesburg. "Most of the people who attended that conference were white, middle-class academics," he recalled, "but at the back of the hall where I gave my talk I noticed a small group of shabbily dressed black men wearing what looked like Jewish skullcaps. When I spoke to them, they announced with pride that they were members of the Lemba Tribe. They said that they were Jews and that they'd come from the Middle East centuries if not millennia before. I found this rather intriguing but very difficult to believe. They didn't look Jewish, and nobody at that time knew that there was any sort of Jewish penetration into black Africa. It seemed absolutely mythic."

Parfitt talked for hours with a young man named Phillemon Matsherry, a clerk by day and a law student by night, who headed the Lemba Cultural Association in Soweto. "I am a Jew, but unfortunately I have a Venda nose, not a Lemba one," he told Parfitt as he rubbed his broad, flat nose. "All my family have fine long noses. I am the only one with this!" Matsherry related a remarkable tale of the Lemba's history and traditions, including the claim that his Jewish ancestors designed and built the Great Zimbabwe, which translates as "the great stone city," the vast, medieval ruins that gave the country its name and even today is the focus of great political and racial controversy in Africa.

The Lemba are a tiny tribe of about fifty thousand, consisting of a pocket population in Soweto and the remainder dispersed across northern South Africa, the south of Zimbabwe, and in Mozambique. They speak a variety of Bantu languages. Today most are Christian. But according to their oral legend, they are Israelites deported by King Nebuchadnezzar from Judea to southern Arabia, a land they call Sena. The Lemba cite passages in Ezra 2:35 and Nehemiah 7:38, where it says almost 3,930 "sons of Senaah" return from exile in 538 BCE at the behest of the Persian king Cyrus to rebuild the Temple the Babylonians had destroyed. The Lemba believe that those banished Hebrews who chose to remain in Sena eventually traveled from the "northern city" and "across the ocean" to the East African coast. From there, they moved inland. Although they constructed an immense and impressive new city, like the biblical Israelites before them, they angered the Lord by eating unclean food. God took retribution, exiling them to live among people who did not share their beliefs.

Parfitt had known about the Great Zimbabwe since he was a young boy, devouring Henry Rider Haggard's page-turner *King Solomon's Mines*, written in 1885, shortly after the German explorer Karl Mauch stumbled across the fantastic ruins. The ruins are seemingly in the middle of nowhere, lying just to the east of the Kalahari Desert. With no credible evidence but a fiery imagination (he mistook local forest wood for Lebanese cedar, further clouding his already-fevered judgment), Mauch declared that the crumbling castle, a honeycomb of intricate buildings and piazzas spread over 1,800 acres with an outer wall 32 feet high and 17 feet thick, was the queen of Sheba's legendary golden city of Ophir. The most impressive surviving structure is a conical tower shaped like a beehive that probably was part of the royal residence. Overflowing with biblical enthusiasm typical of the times, Mauch opined that Sheba was in fact the queen of Zimbabwe and that one of the three Wise Men mentioned in the New Testament as coming to Israel to pay homage at Jesus's birth was from here as well.

There were many theories of the city's origins, often tied to ambiguous biblical references. "Zimbabye is an old Phoenician residence and everything points to [it] being the place from which Hiram fetched his gold," wrote Cecil Rhodes, the British financier who established the colonial empire of Rhodesia, in the late nineteenth century. "[T]he word 'peacocks' in the Bible may be read as 'parrots' and among the stone ornaments from Zimbabye are green parrots, the common kind of that district, for the rest you have gold and ivory, also the fact that Zimbabye is built of hewn stone without mortar."

In colonial South Africa, the Lemba were known as Kruger's Jews because President Paul Kruger, president of the Transvaal in the late nineteenth century, was said to have "discovered" them. The first Jew to open a butcher's shop in the Northern Transvaal would hire only Lemba as slaughtermen, which the Lemba cite as evidence of the purity of their faith. In the 1960s, the Rhodesian government began promoting the local Lemba as the original builders of the Great Zimbabwe, noting their claims in their origin myth and their monotheism, which distinguished them from many neighboring tribes. It's believed the government commissioned the writing of a 1972 book, *The Origin of the Zimbabwean Civilization*, which claimed the Lemba were of Jewish stock. The endorsement of the Lemba's claims by the white regime served its racist purpose: if the Jewish Lemba built the Great Zimbabwe and its successful civilization, that was further "proof" that "pure" Africans were incapable of architectural feats or of governing themselves.

After the end of white rule in 1979, when the nation's name was changed from Rhodesia to Zimbabwe, black Africans gleefully and quite reasonably claimed the once-great city as their own, although the evidence was thin. Anthropologists believe that the finest walls were probably constructed during medieval times, when the city sat strategically on a lucrative gold and ivory trading route leading to the Mozambique coast. But by whom? Because the Shona people, a Bantu tribe, live in the area of the Great Zimbabwe today, some archaeologists speculated that they may have started building the complex in the twelfth or thirteenth century. There are also sketchy tales of a longtime Semitic presence. The first reference to a possible inland empire in the area of the Great Zimbabwe is the gold-rich ancient kingdom of the Sabeans, a non-African people from the Near East, mentioned in 1 Kings 10 and cited later by Pliny the Elder around 70 CE. There were also scattered reports in the sixth and tenth centuries of an active Arab gold trade in the region.

White colonialists were (and often are today) ready to embrace any theory that assumed the Great Zimbabwe was built by anyone but indigenous Africans. "They are baboons; they do not build anything—they destroy," Parfitt quotes one white Zimbabwean woman as telling him. Parfitt says he was at first almost certain that the massive edifice was solely an African enterprise. Nonetheless, and although he was disgusted by the dismissive comments of white racists and concerned that their ravings might gain credibility if the Lemba's claims of being of Semitic origin and the builders of the medieval city proved true, he was intrigued by their story. At that time, the small tribe was on no one's radar screen—"When I started my work on the Lemba, no anthropologists that I knew had even heard of them," Parfitt told me. "They weren't even listed in the African dictionary." But he became hooked immediately.

One day after his unexpected and intriguing encounter with the young Lemba law student, Parfitt took off for the Venda homeland in the northeast corner of South Africa. His views of the claims of the Lemba changed almost immediately. "In the course of the weekend, I could see that it was almost certain that they must have some kind of Semitic connection, because all of their premodern religious and social practices seemed to be imbued with a quality that was essentially Middle Eastern, essentially Semitic."

Classic Jewish stereotypes have been layered upon the Lemba over the years: they were considered industrious and good at business, and had a devotion to education. Their metalwork was considered far superior to that of the surrounding tribes, and until recently, they were known to build with stone without using cement, echoing the unique style found in Great Zimbabwe. The Lemba practice strict food rituals as laid down in Leviticus 11. They eat no meat that has not been ritually slaughtered kosher-style by a tribesman, and they do not eat the meat of pork and piglike animals, such as the hippopotamus, which explains the derivation of their Bantu name, Lemba (also Balemba, MaLemba, or Varemba), which means "people who refuse." Their many Jewish-like practices include circumcision; the prolific use of biblical names like Solomon and clan names—Hamesi, Sadiki, Sulamani—that sound vaguely Hebraic; and a taboo on intermarriage by Lemba men with women from other tribes. They call non-Lemba *wasenzhi*, "the gentiles," although non-Lemba women may marry into the tribe if they convert. They often converse in a secret language, Hiberu, meaning "Hebrew" in the Shona language. The Star of David and "elephant of Judea" decorate their homes. Physically, they stand apart from many of their neighbors: unlike the Shona, many are lighter-skinned and have aquiline noses and narrow, non-Negroid lips.

The few scholars who even had heard of the Lemba were dismissive of their claims. They were considered Judaizers, much like the Falasha. As recently as 1997, Aviton Ruwitah, the senior curator of ethnography at the Museum of Human Sciences in Harare, had written an impressive paper decimating the Lemba's folk belief in their Israelite roots. He noted that the Star of David, the favored symbol of the Lemba, did not come into widespread use as a symbol of Judaism until the Middle Ages and therefore could not have been brought to Africa thousands of years earlier. Their secret language, Hiberu, has its clearest roots not in Hebrew but in Shona. And many of their cleanliness and food rituals draw as much from Islam as from early Judaism.

Skeptical but curious, Parfitt spent months trekking by train, car, and boat from South Africa to Zanzibar looking for any corroborating evidence of the Lemba's ancestral claims. "Once you get to like the people that you're working with, and I got to like quite a number of the Lemba a good deal, you really want in some curious way to believe what they have to say about themselves," Parfitt told me. But his research turned up nothing to challenge the prevailing anthropological views. "I was really in the curious position of getting fewer and fewer indications that this group was of Jewish origin," he said.

In 1992, Parfitt published his story, *Journey to the Vanished City: The Search for a Lost Tribe of Israel*, and reached a conclusion in line with almost every other anthropologist: the Lemba had conflated their local folk and religious myths with fragments of Islam and slivers of tales shared with them by Christian missionaries and Jewish merchants. In other words, he believed the Lemba, like other Jewish-like African sects, were well-meaning fakers, victims of their naïveté.

Then the DNA detective work began.

GENETIC WITNESS

One thread from Parfitt's original research remained loose: the mystery of Sena. The Lemba constantly talked of a coastal city that medieval Arab geographers knew as Sayuna, which is pronounced much like "Zion." None of the Lemba leaders knew for sure where it was. "It acts as place of origin but also the place to which they go," said Parfitt. "They refer to it in the same way that we would refer to paradise or heaven. They say, 'We'll meet again in Sena' and things of that sort. It seemed to me that the whole story was magical."

Some Lemba believe there was an original Sena in Jericho and perhaps one other in East Africa. Were there more? "It was only later when I was doing a book on the Jews of the Yemen that I picked up the trail [that led to Sena] in southern Arabia," Parfitt said. He traveled to Yemen, in southern Arabia, in 1996. The country is home to one of the oldest surviving Jewish communities in the world, although it shrank to almost nothing after the Israeli War of Independence heightened tensions and forced its evacuation. Between 1948 and 1951, in Operation Magic Carpet, an entire community of devout Yemenite Jews was airlifted to Israel, where they were immediately affirmed as citizens under the Law of Return. Like the Lemba and the Falasha, they claim their origin traces to the days of King Solomon, a story they may have borrowed from Ethiopians, who conquered and converted many pagans to Christianity in the sixth century. Crypto-Jews were allowed to resume their religious practices after the Muslims took control of southern Arabia in the eighth century.

Yemenite Jews, who mostly live in suburbs around Tel Aviv, still practice an undiluted form of Judaism. Although they think of themselves as fiercely separatist, claiming to have lived for centuries in tightly knit Jewish ghettoes, their genes tell a more ambiguous story. Like the Falasha, they share more genes with their non-Jewish neighbors than do most other Jewish communities around

the world. Some genes for blood groups and enzymes common in
Yemenites are not found in other Jewish groups, but are frequent in
Arabs. Moreover, they don't suffer from many of the so-called Jew-
ish diseases that affect nine-tenths of world Jewry. The genes of the
Yemenite Jews suggest they have often intermixed and welcomed
converts.

While traveling through Yemen, Parfitt came across an imam
in the holy town of Terim in the Hadramaut, a harsh desert valley known as the "courts of death." "I had been thinking that the
Lemba's Sena might be Sanaa, the capital of Yemen, or some other
place, possibly even Sayhut, which is one of the adjoining towns
that sounds rather like Sena. He said, 'But there's actually a place at
the end of the Wadi Masilah that is called Sena to this day.'"

Parfitt was flabbergasted. He compared the Lemba's folk accounts with the geography, wind charts, and tides that link southern Arabia with the East African coast. During monsoon season,
the winds blew down the coast of Africa, and for the rest of the
year, the winds reversed. Perhaps Sena lay at the northern end of
the trade route, in the Yemeni desert. Suddenly, their story seemed
plausible. But it was still just anthropological guesswork.

DNA would confirm the calculus. Using classic genetic markers—blood groups, enzymes, and the like—the Lemba tested out to
be no different from their African tribal neighbors. But that would
be expected if, as the Lemba claim, they were Semites who migrated from Sena to Africa thousands of years ago and took foreign
wives. If their folk myth was true, the evidence would not likely be
found in classical markers, but they should show up in the male
chromosome—their male lineage.

Unbeknownst to Parfitt, a fellow Welshman, Trefor Jenkins at
the University of the Witwatersrand in South Africa, had hatched
a study of his own to test that very theory. Jenkins and his colleague Mandy Spurdle swabbed about fifty Lemba men, then published their study in the *American Journal of Human Genetics* in
1996. They concluded that approximately 50 percent of the Lemba
Y chromosomes are Semitic in origin and 40 percent are African,

while the rest cannot be placed. Two subsequent female DNA studies, one coauthored by Jenkins, found no evidence that the Lemba
women were of Semitic origin. In other words, based on these ancestral markers, many of the male ancestors of the Lemba came
from outside of Africa but had children with local women upon
their arrival. The researchers had definitively showed that at least
part of the Lemba's oral tradition was correct. But Semitic could
mean Hebrews, Arabs, Assyrians, Phoenicians, or others. Were the
men descended from Jews?

The Lemba story was first brought to the public's attention in
the 1996 BBC television series *Origins*, written and narrated by the
University College London geneticist Steve Jones, who also produced a book, *In the Blood: God, Genes and Destiny*. Perhaps influenced by the skeptical anthropological dogma that first ensnared
Parfitt, Jones went far beyond the ambiguous DNA evidence to
pooh-pooh the Lemba's claim of Jewish origins:

> The tale of how those genes reached Africa is more prosaic than
> that of a band of exiles fighting their way across the continent
> to build a great temple. The town of Sayuna in Mozambique
> was a center for Arab traders (many setting out from Sana'a) for
> hundreds of years. Africans—including the ancestors of today's
> Lemba—traded there and learned Moorish customs, circumci-
> sion and ritual slaughter included. The Arabs took wives from
> the coastal tribes. Their children were brought up in Islam. In
> the sixteenth century, Sayuna was destroyed by the Portuguese,
> and its Arabic culture disappeared. The Lemba, though, retained
> memories of Solomon and Moses, both of whom are as impor-
> tant in Islamic as in Judaic belief. Rather like the Falasha of
> Ethiopia (whose Judaism may derive from an attenuated form
> of early Christianity stripped of its New Testament elements)
> Lemba Jewishness stems in part from a diluted and almost for-
> gotten Islam.

Notwithstanding Jones's authoritative-sounding speculation that
the Lemba originated as Arab traders, the riddle of their origins
had not been solved. So Parfitt was at least a little curious when

he received a phone call out of the blue from Neil Bradman at the Center for Genetic Anthropology in London. Over lunch, Bradman outlined his bold project to map Jewish history in the genes, starting in Africa. Collecting samples in Jerusalem and London was one thing; venturing into rural Africa was another challenge altogether. The Jewish millionaire businessman turned geneticist was just too intimidated to go into the bush alone to collect DNA samples. He hoped Parfitt could help him.

"He started the process of trying to explain to me what genetics was all about," said Parfitt. "And I realized that I had not the slightest interest. This was a whole lot of complete gobbledygook as far as I was concerned. But then I was seduced by Neil."

Renowned for his charming persistence, Bradman was not about to take no for an answer. A short time later, he and Parfitt were on a plane together bound for Ethiopia to collect DNA from the Falasha and the Qemant, who live in the Lake Tana region and also claim Jewish ancestry. Then it was on to Kenya and Uganda. As he came to appreciate the potential power of gene mapping, Parfitt prevailed on Bradman to take samples from the Lemba as well. Bradman and Parfitt were not optimistic that the DNA they would collect could resolve the mystery. At best, they believed, they would end up with a map of the migrations of the Lemba—an interesting enough story on its own, considering the Sena link in Yemen that Parfitt had stumbled upon.

Parfitt took two trips to Lemba villages in 1997, collecting DNA samples from twelve clans. His first stop was a black township in Vendaland, South Africa. The Lemba leadership, who by this time knew and trusted Parfitt intimately and was aware of the results of the Jenkins study, welcomed him warmly. "From their standpoint," Parfitt recalls, "this research might confirm beyond any possible doubt what they thought about themselves—and what they wanted to think about themselves—that they were of Jewish origin." He filled a notebook with the details of each person's age, place of birth, paternal relations, and clan. Then the samples were shipped back to London for analysis.

Bradman and Parfitt also flew to Yemen to try to secure some samples from Muslim men to test the Sena theory. Their first stop was in Terim, where they addressed a quizzical audience of some seventy students. After Parfitt explained what they were trying to do, one fundamentalist student stood up to condemn the project as a violation of the Qur'an (often called the Koran in English)—no passages were cited—and led a walkout of all but two, avowedly secular, teenagers. The two students allowed their cheeks to be swabbed and were awarded by a relieved Bradman, who feared the entire trip was going to be a bust, with color Polaroid snapshots of the magic moment. "Brandishing their spoil, they marched out into the blistering heat of the yard," Parfitt writes. "These were probably the first color photographs of themselves they had ever possessed. Within a minute the rest of the students were queuing up to give their DNA and to receive their gift." Overall, they collected almost four hundred samples.

The first results indicated that the Muslim Yemenites from the Hadramaut and the Lemba share a common ancestry. That was not entirely unexpected, because many Lemba clan names correlate with place names from this desert region in Yemen. What about the Lemba's claimed Jewish origins? As the lab results began to trickle in, the excitement grew at the Center for Genetic Anthropology in London. The Lemba men showed a mixture of Semitic and Bantu markers at roughly a two-to-one ratio. Almost 9 percent of all Lemba males tested carried the Cohen Modal Haplotype, two to three times that of most Jewish populations. But there was even more gold in this data.

"Neil rang me up on the telephone," Parfitt remembered. "He was quite amazed. One of the Lemba clans, a group I had not paid much attention to up to this point, the Buba, carried the Cohen haplotype at percentages as high as Ashkenazi Jewish Cohanim. We were all quite amazed, frankly." Although the CMH is almost nonexistent in Yemenites today, more than half of the Buba clan, 53 percent, had the priestly signature marker. Here was genetic evidence that the Lemba were of Semitic ancestry and most likely descendants of priestly Jews.

According to Lemba tradition, the Buba are descendants of a man named Buba who led the original tribal members out of Judea. They are the Lemba priests for ritual purposes and the senior clan. "'Buba' means 'Judea,'" says Ephraim Selamolela, a rich Buba from the northern region of South Africa who goes by the name Sela, which means "rock" in Hebrew. "We're the same as the other Lemba. But our noses and skin [color] are different. The Buba also didn't intermarry much. We were marrying with cousins. A sort of royal family." The spiritual head of the Lemba in South Africa, Professor Matshaya Mathiva of the University of the North, is a Buba.

The Lemba study, published in February 2000, directly assaulted anthropological orthodoxy. "It was one of the first times when I fully appreciated just how powerful a tool we had in our hands," recalled Bradman, savoring the recollection. "I started to believe it might just be possible to work up a complete genetic history of the Jews, all the way back to biblical times."

LOST TRIBES AROUND THE WORLD

Over the past decade, geneticists have begun combing through the DNA of far-flung ethnic groups and tribes that practice Jewish or Israelite rituals in search of common ancestors. The Kurds of the Fertile Crescent, what is today Iraq, Iran, Syria, and Armenia, have attracted particular interest because of the historical links between Jewish and non-Jewish Kurds. Assyria was supposedly the central home-in-exile for the Ten Tribes. Could Kurdish Jews, seven thousand of whom were brought to Israel between 1948 and 1950, be a remnant of a Lost Tribe? Or are they descendants of pagan converts, as legend has it? It's believed by some historians that when the royal house of Adiabene in ancient Kurdistan converted to Judaism two thousand years ago, great numbers of the Kurdish people may have converted as well. Is any of this history confirmed in the genes?

A DNA study of the male lineage by an international team led by Ariella Oppenheim of the Hebrew University, published in 2001, has put the Adiabene myth to rest. If the legend is true, the geneticists concluded, it "does not appear to have had considerable effect on the Y chromosome pool of the Kurdish Jews." The researchers also found that Arab Kurds and Jews have only distant genetic links. There was "negligible" evidence of any mixing between Kurdish Jews and their Arab former host population, which carries a distinct marker common among Arab populations, including Palestinian Arabs and Bedouins. However, the genes of non-Arab populations living in the northern part of the region tell a different story. By and large, Jews are close blood cousins to Christian Kurds, Armenians, and Turks.

Legends abound about Lost Tribes settling in India and Asia, but only a few myths have yet been tested using DNA technology; for the most part, the evidence for exilic Jewish communities is little more than speculation fueled by Christian seekers. For example, some Christians believe that almost all Asians are of Semitic ancestry. That myth can be traced to a nineteenth-century Scottish missionary, Norman McLeod, who suggested that Asians descended from Noah's son Shem (which, he claimed, made them Semites), who escaped to the east during the Assyrian debacle. The Israelites supposedly wandered all the way to Korea, where some split off for China and others for Japan. What was his proof? The sheep he came across in Asia looked eerily like the breed of sheep imported from Palestine and sold in Smithfield market in London!

Although there is no historical evidence to support their beliefs, a surprising number of Japanese claim they are descendants of Lost Tribes. Proponents of this view cite the fact that the Japanese use scrolls, as did the ancient Israelites, and have eaten unleavened bread since ancient times. The leadership of the Yamato clan claim they have Israelite roots, while the Hada believe they are children of

Zebulun. Some Japanese aver that the colorful festival of Gion held each June in the ancient capital of Kyoto began as a celebration of the promised return to Zion, noting an uncanny resemblance between the portable shrine carried in the festival and ones used by the Hebrews in their long sojourn to the Promised Land—how that's known, exactly, is not clear. Many Makoya, a Zionist Christian sect known for their fondness for temples, assumed they inherited their passion from the temple-building Hebrews, of course. Members often take on Hebrew names and follow Jewish observances, although there is no archaeological or genetic evidence to support their claims of Israelite descent.

Perhaps the most elaborate legend, most famously promoted by a Canadian professor and columnist, Edward Odlum, during the 1930s, and now circulated on the Internet, revolves around the lost tribe of Menasseh. Professor Odlum explained that Menasseh rulers headed a group of Israelite tribes that traveled eastward during the Assyrian Exile, settling in Japan. He believed that they were half Egyptian, which in his mind explains why Japanese eye and skin tone is so different from that of their close neighbors, the Chinese. But there's more. The warriors of Menasseh were undoubtedly heavily involved in the defense of Samaria. When they resettled in the Japanese mountains, they became known as samurai. What could be clearer? Unfortunately, the DNA evidence suggests that the Japanese are of exclusively Asian ancestry, with no Semitic markers.

Lost Tribes have supposedly been cited throughout China. Szechuan Province, which was a stop on the famous Silk Road trading route, is often mentioned as a gateway in many Asian Lost Tribe stories. There is some evidence that small bands of Jews settled in China around 240 BCE. There are historical references to a tiny Jewish trading enclave in the dynastic capital as far back as the seventh century CE and a larger influx of Jewish traders during the eleventh and twelfth centuries.

When China was finally opened to outsiders in 1840, missionaries found a large community, the Chiang (or Chiang-Min), in the remote northwest of Szechuan Province. They claimed to be descendants of Abraham, perhaps through the northern tribe of Menasseh, whose ancestors had fled the Assyrians, heading first to Afghanistan and then to Tibet, where they settled along the border. The Chiang later caught the fancy of Reverend Thomas Torrance. The Scot's 1937 book, *China's Ancient Israelites*, expounded his belief that the Chiang-Min were really lost Israelites. After all, he noted, they mounted twelve flags (for the Twelve Tribes?) beside the altar, called on a god named Y'wa when times were bad, and used an ancient-Israelite-like plow drawn by two oxen, as stipulated in the Bible. Citing Paul's declaration in Romans 1:16 to bring the Gospel "to the Jew first," Torrance aggressively tried to convert them to Christianity. About 250,000 Chiang still live in fortlike villages in the high mountain ranges. Among their more curious ancient-Israelite-like rituals, the Chiang sprinkle blood on doorposts to ensure the safekeeping of the house. To date, the Chiang-Min have not undergone any DNA tests.

The largest "Jewish" settlement in China was in Kaifeng, the capital of Hunan Province, in eastern China. The small Jewish community enjoyed the protection of the Chinese rulers, prospering while maintaining their ancestral customs. Though they dressed like the Chinese and spoke Chinese, they prayed in Hebrew. In 1163, a new synagogue was constructed in Kaifeng and renovated in the fifteenth century, both times at government expense. Though the synagogue remains standing today, there are fewer than one hundred Chinese who claim to be Jewish, not enough people for a comprehensive DNA evaluation. If the few remaining Chinese Jews are of Semitic origin—and many do have facial features vaguely suggestive of Semitic ancestry—they are more than likely descendants of traders who traveled the Silk Road and married local women.

There are two Jewish-like communities in India: the Cochin Jews of South India and the Bene Israel of West India. Both claim they came to India as exiled Israelites. The Cochin believe they are descendants of Jews who arrived after the destruction of the

Second Temple. They have long seen themselves as being racially distinct and superior; they turn out a disproportionate number of doctors, lawyers, university professors, and other accomplished professionals, which they attribute to their Israelite ancestry. In 1948, the community, numbering some 2,500, was resettled in Israel. Fewer than one hundred Cochin Jews remain in India today, living a twilight existence near the only functioning synagogue. DNA research out of the Center for Genetic Anthropology suggests that Cochin Jews may indeed be descendants of ancient Semites, although classic genetic markers indicate that there has been a lot of intermarriage over the centuries.

The Bene Israel, or "Sons of Israel," is India's best-known Jewish community. Numbering about four thousand and concentrated mostly in Thane near Mumbai, the Bene Israel claim to be related to the ancient Israelites, but they fiercely reject being called Jews, because they believe they are not descendants of Judeans. According to tribal legend, they are Galilean descendants of the northern tribes from the Israelite territory north of Samaria who escaped Palestine during the reign of the Greek tyrant Antiochus, around 175 BCE. They are said to have fled to Egypt, where they boarded a ship bound to India; the ship never made it. All of their possessions, including their Torahs and prayer books, were supposedly lost at sea, and only seven shipwrecked couples managed to swim to safety.

Unlike the Cochin, the Bene Israel have had little or no contact with Jews over the centuries, so the origins of their practices have long been a mystery. It's plausible they originated as primitive Bedouin Jewish tribes or perhaps came to India as traders escaping the periodic uprisings that erupted in the East during the first millennium. A nineteenth-century Jewish traveler named Israel ben Joseph Benjamin had no doubt the Bene Israel he encountered were lost Jews. He was convinced that the Ganges River was none other than the river Gozan mentioned in the Hebrew Bible.

After his remarkable investigations of African Jews, Tudor Parfitt turned his attention to the Bene Israel. At first, he believed their

suspected she might be of crypto-Jewish origin. When she was diagnosed, she sought counseling from a rabbi in Colorado. "You're Jewish," the rabbi told her. "It doesn't make any difference if you are an atheist, you are Jewish. You don't have to convert. You can just start practicing the laws if that's what you want to do."

The intellectual shoot-out between Stanley Hordes and Judith Neulander over the origins of these "pocket Jewish" communities seemed to have been settled in Hordes's favor when, in 2006, Neulander participated in a new study with a distinguished group of anthropologists and geneticists. They randomly sampled 139 men of Spanish ancestry from across northern New Mexico and southern Colorado and compared mutations on their Y chromosome to those of Spaniards. Their markers were genetically indistinguishable and very different from Jews, both Ashkenazim and Sephardim. "Although Spanish-Americans undoubtedly have some Jewish ancestry, they appear to have no more than do Iberians," the researchers wrote.

Neulander's findings and Hordes's conclusions, supported by a slew of Jewish mutations found in Hispanos, appear to contradict. Who is right?

Actually, their research does not necessarily contradict. Neulander addressed the question of whether a "substantial" number of crypto-Jews settled in what is now New Mexico in the late sixteenth century and concluded based on the mutation study that they did not. But Hordes had never made that claim, saying only that the evidence suggests at least some crypto-Jews were among those early settlers. One key flaw in the study: it didn't appear to include self-identified crypto-Jews, who are scattered randomly in small numbers through the Hispano population. The unusual findings of "Jewish disease" genes or Jewish priestly markers in some Hispano communities could be unexplained aberrations. More likely, there are pockets of descendants of former Jewish families in various Christian Hispano communities.

I asked Mary-Claire King, who has also come across Mexican American women carrying the deadly Jewish mutation on BRCA1,

what to make of this dispute. "The 185delAG allele among Hispanos of the American Southwest is the Jewish allele. No question
of this genetically," she wrote me. "It's on the same ancient haplotype as in undisputed Jewish families. Of the three Jewish BRCA1
and BRCA2 alleles, BRCA1.185delAG is the oldest and has been
observed (in Israel) in Iraqi Jews as well as among the Ashkenazim.
Therefore the Hispanos could be originally Iraqi Jewish or Sephardic or Ashkenazi. Of course how they define themselves culturally may be different. But genetically, at least one allele of their
ancestry is Jewish."

The Jewish disease mutations found among many Christian
Hispanos practicing Jewish-like cultural traditions does suggest
another intriguing conclusion: many crypto-Jews married other
crypto-Jews—they were endogamous, just like their Jewish cousins.
That's what appears to have happened with the extended Sánchez
clan, who speak openly about many of their Jewish-like rituals.

Working with Family Tree DNA, Father Sánchez has initiated
his own study. While Neulander and her coauthors found the Jewish priestly Cohan Modal Haplotype at very low frequency in the
small number of Spanish Americans they tested, the Santa Fe Project, as it's called, indicates that some Hispano clans show an unusually high frequency of the CMH. Sánchez purportedly traces his
genealogy back to the Carvajal family. His DNA—a broad analysis
of a wide range of markers known to mark genetic roots—showed
73 percent European ancestry, most of it matching with Sephardic
Jews; his father's tested at 81 percent, again mostly Sephardic.

For those who claim descent from crypto-Jews, the academic
dispute is beside the point. They are emboldened by what they
believe is a richer understanding of their Israelite roots. Although
some Hispanos who consider themselves crypto-Jews have converted, most of them still practice Catholicism, but with a twist.
They participate in Jewish rituals and study Jewish history. "It's a
weird product of the Inquisition," says Daniel Yocum, a member
of Sephardim for Yeshua, a messianic sect of one hundred members of Adat Yeshua, an Albuquerque synagogue that holds regular

Sabbath services for those who consider themselves crypto-Jews. "In no way do we want to abandon Jesus Christ the Messiah. But considering all the Sephardim went through, it wouldn't be right to abandon [Judaism] outright . . . to let them die off is almost cruel."

Beatrice Wright is now convinced that she too is of Jewish lineage. "At first, I was shocked by the findings," but as she thought back to her childhood, she remembered her suspicions that many Hispano families seemed different from other Mexican Americans living in the valley. They lit candles on Friday evenings and adorned the tombstones of relatives with the Star of David. Wright recalls reading a novel about a Jewish girl whose family always swept dirt to the middle of the room because they did not want to defile the door mezuzah—the tiny scroll of parchment containing biblical passages placed in a small case and affixed to the doorpost in Jewish homes on the right side as you enter to serve as a reminder of God's presence. Although her family did not have a mezuzah, she remembered that the women in her extended family avoided sweeping past the doorways: "When we talked about it, we all laughed and said, 'We're probably Jewish, too.'"

She clearly understands the fateful consequences of the finding for her extended family. There are now dozens of potential victims with no idea of their vulnerability. She has put together an information packet about her family tree that she passes along to newfound relatives to encourage them to go for testing. It includes a photograph of eight of her aunts, half of her father's family of sixteen brothers and sisters—six of whom have fallen victim to ovarian or breast cancer, with only one still alive. One of her uncles also died of prostate cancer, which can be caused by the same mutation.

The reaction by her relatives to the news has been complicated. While most family members have welcomed the warnings and some are delighted to discover hidden nuggets in their ancestry, others have been shocked. Her daughter, when turning thirty and facing a very high risk of having the mutant gene, was the most resistant.

"She's scared. She said, 'Mom, I don't want them to start lopping off my body parts.'" Wright sighed. "My heart aches with the fact that I may have passed this mutation onto my children and their future children. I'm proud of discovering my Jewish heritage, but this is a heavy burden to accompany it."

CHAPTER 9

ASHKENAZIM: CONVERTS OR ABRAHAM'S CHILDREN?

According to the Hebrew Bible, the early Hebrews were a mixture of tribes that coalesced into a people more than 2,500 years ago in the Middle East. Then came a series of exiles and the Jewish diaspora, which scattered Jews into small communities around the world. Are modern Jews the descendants of these Israelites, as Jewish lore holds?

History provides an incomplete answer. The smallest fraction of the modern Jewish population, Jews from the Middle East known as Oriental Jews, can most easily trace their roots to biblical times. As we've learned, Sephardic Jewry, founded by Jews who made their way to Iberia and North Africa over many centuries, has been decimated by mass conversions. The origins of Ashkenazi Jewry are less certain. The thousand years after the destruction of the Second Temple as Jews trickled into Europe are only sketchily preserved in the historical record.

One controversial theory of European Jewry, once widely embraced by historians and the public, holds that most modern Jews are the descendants of converts. The predecessors of Ashkenazi Jews supposedly did not originate in ancient Palestine, but rather in pagan Khazaria, a medieval empire in Eurasia extant from the

KHAZARIA: 7th-10th CENTURY

KHAZAR EMPIRE

Kiev

BULGARIAN
EMPIRE

Sarkel

Itil

Caspian
Sea

CRIMEA

Kherson

Danube R.   Pliska
            Prestav

Black Sea

Caucasus Mts

Serdica

Adrianople 81

Constantinople

MACEDONIA

Rome

BYZANTINE EMPIRE
(EAST ROMAN EMPIRE)

Tarsus

Baghdad

Mediterranean Sea

Damascus

Jerusalem

Alexandria

Cairo

ARABIA

Figure 9.1. Jewish Khazaria?

seventh to the tenth centuries CE. According to Turkish and Jewish lore, in the eighth or ninth century, King Bulan decided it was high time for the royalty to formally adopt a religion. He staged a competition of sorts between the three monotheistic religions. Supposedly, the Jewish representative rhetorically outdueled his rivals. By legend, Bulan chose Judaism, converting the royal family. Bulan's successor, Obadiah, learned the Mishnah and Talmud and strengthened the royal family's ties to Judaism, inviting rabbis into the kingdom and building synagogues. Some stories held that the entire kingdom converted to Judaism.

What does the historical record report? The real story is a mixture of myth and mystery. Some Lost Tribe fabulists believe a tiny founding Jewish population arrived in what became Khazaria as part of the Assyrian Exile recounted in 1 Chronicles and 2 Kings. It is known that the Assyrians had outposts in "Eden," which is thought to be a vast area encompassing modern-day Armenia and the plains north of the Caucasus Mountains. Some Jews did migrate into the region during the seventh century. At that time, the declining population of Mediterranean Jews was caught between the rising tide of Islam, which swept through the Near and Middle East, and Byzantine Christianity. Some Jews banished from Constantinople by Leo III more than likely drifted north into neighboring Khazaria.

The Khazars were a seminomadic people of mixed stock— mostly Turkish, it is assumed—who united to form an independent kingdom. At its most expansive, the Khazarian Empire extended from Byzantium north to deep into Russia and west to Kiev in the Ukraine. The Khazarians came to dominate trade along the Volga and occupied a strategic position. They developed a strong army that buffered Byzantium from the barbarian hordes and sacrificed their blood to block the Arab expansion into Europe. The scanty historical record suggests that the various Khazarian tribes practiced shamanistic and pagan cult worship. King Bulan's religious shoot-out supposedly changed that overnight.

During the early twentieth century, historians speculated that the Mountain Jews of the eastern Caucasus and various Muslim Turkic groups in the North Caucasus might be descended from the Khazars. Beginning in the 1950s, the legend of Jewish Khazaria emerged as a full-fledged theory. In that pregenomic era, some scholars noted that many eastern European Jews have red hair and blue eyes, which are not common Semitic features. They also maintained that Yiddish, the universal language of the Ashkenazim, may have many Germanic words, but is syntactically a Slavic language. From this they cobbled together a new theory of Jewish history: modern Jews are not children of Israel, but descendants of pagan converts from Turkish Eurasia.

At the time, the world was still recovering from the devastation wrought by the racial militarism of the Third Reich and the empire of Japan. The prevailing notion that Jews were a cohesive people with blood ties to the biblical Middle East came uncomfortably close to discredited racial theories. Maybe it would be better, for Jews and history, if Jews weren't such a "pure race." Not surprisingly, the Khazarian conversion theory, even absent significant archaeological or historical evidence, was attractive. It became a favorite of leftist Jews, who viewed with wariness the flood of displaced Jews heading to the newly created State of Israel based on Jewish and Christian fundamentalist claims of a Jewish "right of return" to biblical Zion. They believed that any hint of a Jewish "race" fostered persecution and prejudice—both against Jews and by them against Palestinians. "That the Khazars are the lineal ancestors of Eastern European Jewry is a historical fact . . . though the propagandists of Jewish nationalism belittle it as pro-Arab propaganda," wrote Alfred Lilienthal, a Columbia University professor, in 1953 in his polemical attack on the new Jewish state, *What Price Israel?*

The most influential propagandist for the Khazarian theory was Arthur Koestler, a Budapest-born Jew who survived World War II to immigrate to England. The notion of chosenness, he came to believe, "sets the Jew apart and invites his being set apart." The Khazarian theory was a perfect antidote to Jewish separateness. In his 1976 book *The Thirteenth Tribe*, he argued that as the Khazarian Empire crumbled, remnants of the converted Jews migrated into eastern Europe. They mixed with pockets of Jews already living there to form the seeds of Ashkenazi Jewry. To support his speculation, Koestler assembled historical anecdotes, noting for example that the Hebrew Bible refers to a people known as the Ashkenaz who lived around Armenia. "[T]he majority of . . . world

Jewry might be Khazar and not Semitic," he concluded. "[T]heir ancestors came not from the Jordan but from the Volga, not from Canaan but from the Caucasus, once believed to be the cradle of the Aryan race . . . [G]enetically they are more closely related to the Hun, Uigur, and Magyar tribes than to the seed of Abraham, Isaac, and Jacob."

Because of his status as a renegade intellectual, the Khazar theory soon became the popular wisdom and remains so in some circles today. But this thesis, however progressive in its original intent, has provided fuel for decades of anti-Jewish ramblings and is widely circulated among anti-Israeli Arabs and white Christian supremacists. "Most 'Jews' so-called are at best Gentiles and at worst Serpent's seed as evidenced by their enmity against Adam's race, and with no inheritance in this earth, their international proclivity," rails a typical Internet screed that approvingly cites Koestler's decades-old book. The controversial notion of the right of return has endowed the Khazarian theory with toxic impor-tance. And what a sad irony: while the Nazis damned anyone with even "one drop" of distant Jewish ancestry as "real Jews" and thus indelibly stained, Jew haters quote the liberal icon Ar-thur Koestler as though his words were the embodiment of his-torical truth, thus damning Jews because they supposedly aren't real or pure enough.

Scholars and geneticists are gradually beginning to separate fact from fiction. It's easy to misread the historical documents, says Kevin Brook, an expert on Jewish Khazaria. Brook is an eminently reliable, if academically untraditional, font of information on Khazaria. He is not a formally trained historian but an impressively self-taught scholar, respected around the world and creator of the Web site khazaria.com, which includes a comprehensive collection of studies and journalistic accounts on Jewish ancestry and genet-ics. He's the author of a well-received book, *The Jews of Khazaria*, and has become a legendary figure in Jewish genealogical circles. I first met him years ago at an International Conference on Jewish Genealogy, where academicians and genealogists raptly listened to his lecture on the history of Khazaria. Like many Jewish identity seekers, Brook had little interest in formal religion while growing up and is not a practicing Jew. His passion has always been Jewish culture and genealogy, in particular his family history, which disappears into the hinterlands of the Ukraine in the nineteenth century.

Brook is convinced the religious competition and mass conversion are apocryphal. "Khazar" as it was used in sources properly refers to the ethnicity of the kings and nobles and not to the majority of the populace, he says. "Based on the historical record, only the royal court and select nobility converted." Khazaria had a reputation as a multiethnic empire. Its supreme court reportedly consisted of two Jews, two Christians, two Muslims, and a pagan. Some historians speculate that Khazarian leaders chose Judaism so it could more safely trade with all peoples, surrounded as it was by Christian and Muslim nations.

Whatever the reason for the royal conversion, word of this Jewish-ruled nation reached distant lands, attracting Jews and determined Christian proselytizers. Saint Cyril, a missionary from Constantinople who arrived in Khazaria in 860, reports that he made little headway convincing the Khazarians he encountered practicing a kind of crude Judaism to convert to Christianity. While the tenth-century Muslim scholars Ibn Rustah and Al-Masudi date the royal conversion to around 809, they make no claim that Khazaria had suddenly become a significant Jewish outpost. A century later, however, Abd al-Jabbar ibn Muhammad al-Hamdani wrote that the Khazars as a whole had converted. There is no mention of the conversion by Jewish scholars until the twelfth century, when Rabbi Jehudah ben Barzillai discussed the possibility very skeptically. The story finally gained credence a few decades later, when Judah Halevi, in his book *The Kuzari: A Book of Argument in Defense of a Despised Religion*, sets the conversion date at 740 and controversially imagines the arguments for the supposed spiritual superiority of Jews used by the winning disputant in the shoot-out.

Halevi's account remains a ringing defense of the notion of election. Rather than offering proof of the existence of God, as had the Christian and Muslim presenters, the Jew explained the miracles performed by God on behalf of the chosen people, the Israelites. Religion, he claimed, is less about belief than action. The object of religion is ethical training and the encouragement of good deeds, not to create good intentions. The superiority of Jewish tradition and culture, Halevi's presenter declared, cannot be denied.

There is much doubt whether Halevi drew upon any historical documents when he inscribed his most famous work. It's not even certain that the entire royal family converted to Judaism. "Some members of the Khazar and Alan tribes did voluntarily convert to Judaism," Brook told me. "But for the most part, the Khazar kings guaranteed religious freedom [but did not force the masses to convert]. The Slavs, Goths, Kalizes, Bulgars, and other tribes in Khazaria were pretty much left alone, and they probably kept practicing their old religions, mostly pagan, and only gradually gravitated toward Judaism, Christianity, and Islam. But Judaism did become the state religion, and those who were Jews had an easier time getting government positions. The earliest archaeological evidence of a Jewish presence in Khazaria are silver coins issued around 837 with Arabic lettering bearing phrases like 'Land of the Khazars' and 'Moses is the Prophet of God,' which was supposedly a Jewish version of the Islamic phrase 'Muhammad is the messenger of God.' These coins probably represented the official position of the Khazar rulership that Jews governed Khazaria. However, even a century later, there were members of the royal family who were Muslims."

Because many tribal factions undoubtedly did not embrace Judaism, the size of the Jewish presence can only be estimated. Even at its height, the number of Khazarian Jews probably numbered no more than 30,000 out of a total population of 100,000, including a few thousand nobles and royalty.

It's not clear why or when Jewish Khazaria collapsed. It may

have taken a deathblow with the crushing Russian invasion of
965. There is evidence that a few decades later, a Russian ruler
named Vladimir converted his people in what would have been a
distant province of the empire to Byzantine Christianity. From that
point onward, the Slavs in Russia solidified their links to Byzantium, adopting many cultural and social institutions still practiced
among Slavs today.

By the thirteenth century, the small Khazarian Jewish community had disappeared from recorded history. Most converts probably returned to their pagan traditions or took up Muslim or
Christian worship. Perhaps some retained their identities as Jews.
They may also have mixed with other Jewish converts, notably the
Cumans, also known as the Kipchak Turks, who were known to
be on friendly terms with local Jews in the eleventh and twelfth
centuries. Although there is no evidence of any large Slavic Jewish
community in the early medieval period, there is plenty of evidence
of small Jewish enclaves dotted throughout the Slavic lands, most
notably around Kiev, which could have consisted of descendants
of Khazarian Jews. These villages were probably absorbed by the
waves of European Jews who resettled in Poland and Lithuania
during the centuries of expulsions that began with the Crusades,
in 1095.

The mass-conversion theory, whether of Khazars or other eastern Europeans, remained untestable until the advances in DNA
research techniques in the 1990s. In 1993, in one of the first uses
of Y-chromosomal markers, Italian scientists compared the DNA
of Ashkenazim and Sephardim with non-Jews living in Czecho-
slovakia chosen to represent the possible descendants of gentile
eastern Europeans. If Jews were Slavic converts, the mutations
should be similar. They indeed found a great deal of similarity—but not between Czechs and Jews. The data was startling
for other reasons. For the first time, there was powerful DNA
evidence that Jews from around the world share a common Near
and Middle Eastern ancestry almost untouched by conversion.
The male lineage of the Jews, including those with recent Czech

roots, had far more in common with the Lebanese than with non-Jewish Czechs. The geneticists estimated that the contribution of gentile males to the Ashkenazi gene pool has been very low—1 percent or less per generation.

This was the first concrete genetic evidence that today's Ashkenazi Jews were not mostly converts. What then are their origins and history? How does Jewish lore match with the genetic and historical evidence?

ALL ROADS LEAD FROM ROME?

According to the sketchy historical record, Ashkenazi Jewry probably resulted from a mix of diaspora Jewish communities, including some from Khazaria. But many, if not most, Ashkenazi Jews likely came to Europe via Italy. The Roman Jewish community was established before the Christian era but swelled dramatically when Jerusalem fell into disarray. During the height of the Roman Empire, southern Italy and Sicily emerged as the western center of the postbiblical Jewish world. Jews worked as seamen, craftsmen, and merchants. But by the fifth century, the aging Roman Empire was crumbling under assaults from the Ostrogoths, Huns, Vandals, and other Germanic tribes. In 476, Odoacer the Scirian, the commander and elected king of the German troops in the former Roman Empire, deposed Romulus Augustulus, ending nearly one thousand years of Roman dominance in the Mediterranean. The defeat caused difficult times for gentile and Jew alike, sending many people north into Europe to seek a safer, more stable life.

There are records of Jewish traders venturing from both Italy and the Middle East into Europe and Russia as early as the second century. By the sixth and seventh centuries, Jews were found in Marseille and Cologne and at other Roman commercial outposts in southern France and Germany. By the eighth century, restrictions on Jews in Europe began to ease. The Frankish kings, especially Charlemagne and his successors, actively wooed skilled

Roman merchants, with Jews particularly welcomed. Because they were restricted from owning land, there was no danger that they would become enmeshed in the territorial bickering that plagued feudal societies. Hundreds, perhaps thousands, of Jews settled in Provence, Alsace, and along the Rhine in Cologne, Mainz, Worms, and Speyer, actively trading in swords, furs, and even slaves to the Muslim world, and spices and perfumes to India and China. By the tenth century, Jews were established in northern Europe, and they followed the Norman Conquest, in 1066, into England. Jewish urban enclaves swelled with immigrants, who became craftsmen, artisans, and moneylenders. The Rhineland region, which the Jews called *Ashkenaz*, emerged as a spiritual center of Judaism.

The Crusades, launched in 1095 by Pope Urban II, brought this period of modest prosperity for Jews to a bloody halt. The pope was mostly concerned with retarding the tide of Islamic expansion, but Jews and other infidels were targets as well. Crusaders swept into Jerusalem, killing many Muslims and Jews, including a band of refugees holed up in a synagogue that was burned down. The Byzantine edict forbidding Jews to live in the city was revived, although Jews continued to live in the coastal cities of Palestine. Jews in Europe fared little better. In May 1096, the local bishop of Worms offered the eight hundred Jews of his community the choice of conversion or death; they chose death. The tragedy was soon repeated throughout the Rhineland, and thousands of Jews were reportedly murdered.

Shortly after the first wave of violence sparked by the Crusades began to subside, Rashi of Troyes issued a decree allowing those forced to convert for fear of death to be able to return to Judaism, which reignited tensions with Christians already inflamed by allegations that Jewish moneylenders charged usurious interest rates. In 1179, the Third Council of the Lateran prohibited Jews from having Christian employees, banned Christians from living in Jewish neighborhoods, declared that testimony by a Christian in legal disputes was more reliable than that of a Jew, and banned Christians from moneylending.

With craft guilds closed to them by law and custom and with few other options outside of junk collecting or other menial labors, Jews turned in droves to usury and pawnbroking, but at the price of a stigma that has yet to be erased. According to the Torah, Jews are strictly banned from usury—but only among themselves. "If you lend money to My people, to the poor among you, do not act toward them as a creditor: exact no interest from them," reads Exodus 22:24, an edict reinforced in Leviticus 25:36. However, lending to gentiles is another matter. "You shall not deduct interest from loans to our countrymen, whether in money or food or anything else that can be deducted as interest; but you may deduct interest from loans to foreigners," reads Deuteronomy 23:20–22. Ironically, the directives of the Hebrew Bible, originally written to protect the poor communities of biblical Israel from the financial predation typical of the ancient world, helped establish the stereotype that Jews are obsessed by money. Moneylending jobs would define medieval Jewry, and as we shall later learn, help create the cultural traditions that fostered the high IQ of Ashkenazi Jews.

Although Jews clearly provided a necessary service to Christians, who were banned from lending money, the downward spiral in relations intensified. At the Fourth Lateran Council in 1215, Pope Innocent III decreed that Jews (and Muslims) should be officially branded, "marked off in the eyes of the public from other peoples through the character of their dress." Within a few years, King Henry III of England ordered Jews to wear "the two tables of the Ten Commandments made of white linen or parchment." Louis IX later decreed that all Jews should wear on the front and back of their garments "round pieces of yellow felt or linen, a palm long and four fingers wide." The use of badges became relatively widespread throughout Europe within a couple of centuries and continued to be used as a stigma. It persisted until the Enlightenment, only to be resurrected as the infamous yellow Star of David by the Nazis.

The pope also maneuvered to prevent those who had converted to Christianity from returning to Judaism, as many had done. He issued a papal bull declaring, ". . . he who is led to Christianity by violence, by fear and by torture, and who received the sacrament of baptism to avoid harm receives indeed the stamp of Christianity . . . [and] must be duly constrained to abide by the faith [he] had accepted by force." Innocent also established as dogma the blood ritual of transubstantiation, which assumed that the wine and the thin wafer of unleavened bread consumed by Jesus's Disciples at his Passover Last Supper, and thereafter ceremonially consumed during Mass, are mystically transformed into the actual blood and body of Jesus.

It took only a few years after the papal edict on transubstantiation before Jews were targeted with infamous blood libel attacks. They were regularly accused of stealing wafers, which under the cult of the Eucharist were literally the body of Jesus. The first blood libel massacre erupted in Norwich, England, in 1144, when Jews were accused of capturing a Christian child named William on Good Friday, and in a reenactment of the crucifixion of Jesus, tying him to a cross, stabbing his head to simulate Jesus's crown of thorns, and then using his blood to make Passover matzoh. It was a total fabrication, no solace for the Jews of Norwich, who were massacred.

The blood libel was never given wide credence among educated Christians and was condemned by at least five popes, but the myths about cultic practices persisted in the dark underground of anti-Jewish lore, periodically resurfacing along with accusations, trials, and executions. In most cases, the libel served as a proxy for an attack on Judaism itself. Christians began perceiving Jews as stubborn adherents to an archaic and dying religion. By the thirteenth century, many Catholic clerics had declared the Talmud a forbidden work. They hoped in vain that Jews might be more amenable to converting to Christianity if the Old Testament were their only learned book. In an infamous incident in Paris in 1240, the first of more than a dozen in Europe, twenty-four wagonloads of hand-written volumes of the Talmud were burned as blasphemous after a trial ordered by King Louis IX.

Beset by peasant uprisings and seeing the Jews as easy scape-goats, both France and England expelled their Jewish popula-tions. In 1270, Edward I of England declared all debts to Jews void, which left many of them penniless, then kicked the Jews out altogether in 1290. Philip the Fair of France expropriated Jewish property in 1306 before expelling them. Over the following two centuries, expulsions spread to the Jewish heartland in the Rhine-land. Jews were legislated out of hundreds of towns in Germany, Hungary, and Austria. Some Jews resettled in Provence, in Avignon in particular, which periodically was a magnet for persecuted Jews until anti-Jewish riots in the early fourteenth century. After Catho-lic France absorbed the region in 1481, the Jews of Provence con-verted or were forced into exile.

JEWISH BOTTLENECK

The Jewish population of eastern and central Europe soon sunk to historic lows. Expulsions and murderous pogroms were sand-wiched around the great catastrophe of the time, the Black Death. The bubonic plague wreaked destruction in the Near East before spreading to Europe, wiping out a third or more of its population between 1348 and 1351. Jews were often blamed for spreading the disease by poisoning wells and were often tortured until they confessed their complicity. Pope Clement IV moved to quash the absurd charges, blaming the scourge on the devil in a papal decree, but to no avail. The inflamed peasantry ignored similar edicts by secular rulers. Thousands of Jews were murdered, many burned at the stake, and hundreds of Jewish communities in southern and central Europe were uprooted. Decimated by the Crusades and the Black Death, pariahs in their longtime homes in France and

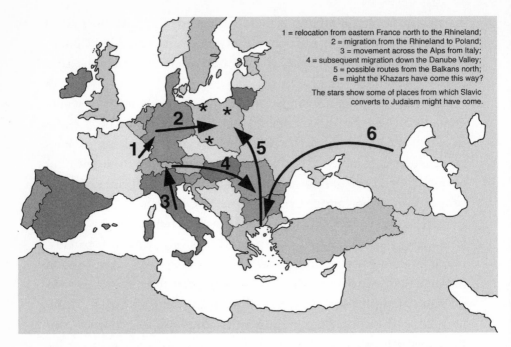

1 = relocation from eastern France north to the Rhineland;
2 = migration from the Rhineland to Poland;
3 = movement across the Alps from Italy;
4 = subsequent migration down the Danube Valley;
5 = possible routes from the Balkans north;
6 = might the Khazars have come this way?

The stars show some of places from which Slavic
converts to Judaism might have come.

Figure 9.2. The origins of Ashkenazi Jewry.

Germany, the Jews of Europe were driven eastward, into Bohemia and Moravia, Lithuania, and the Kingdom of Poland.

Sixteenth-century Europe was the *respublica christeniana*—the universal world of Christendom—dominated by the unyielding sword of the monarchies and under the thumb of a dogmatic papacy. Jews did not fare well. The pope initiated the Roman Inquisition, which was imposed throughout Italy and southern France. In western Europe after 1520, only scattered pockets of Jews remained, and the once-sizable Jewish communities of the German empire continued to shrivel until the end of the century. In the east, the feudal Polish and Lithuanian kingdoms allowed Ashkenazi Jews a measure of autonomy, but it was strictly circumscribed. By some estimates, by the early sixteenth century, the community of European Jews had shriveled to only tens of thousands.

Fleeing from persecution, their ancient religion on the verge of extinction, the People of the Book began turning prejudice to their advantage. Wherever they settled, they brought with them forms of communal organization. They developed their own language, the mélange of Slavic tongues, German, and Hebrew known as Yiddish that became the lingua franca of the wandering Jew. Segregated villages sprang up, ruled by Jewish religious leaders who staked their legitimacy on the Halakhah—the Jewish law interpreted by rabbis. Religion and education became the focus of everyday life.

The central themes of Ashkenazi Jewry—separatism and a devotion to religious literacy—played a profound role in the genetic makeup of Jews. "The more they were oppressed, the more they increased and spread out," predicts Exodus 1:12. And that's exactly what happened to the Jewish population. There is ample evidence in Ashkenazi history of periodic bottlenecks, when the Jewish population shrank, followed by fast growth: about fifty generations ago, around the ninth century, in the Rhine Valley, where Ashkenazi Jewry took root; in the twelfth century, when Jews expelled from southern and western Europe migrated eastward; and in the 1600s, after the Thirty Years' War and the Chmielnicki Massacres that followed.

The most recent bottleneck occurred from 1648 to 1649. The Cossacks, the Orthodox Ukrainians, and the Polish peasantry violently revolted against Polish nobility and the Catholic nobles, but unleashed a special wrath against Jews, some of whom had acted as tax-collecting middlemen. The head of the Cossacks, Bohdan Chmielnicki, claimed that the Poles had sold Cossacks "into the hands of the accursed Jews," a reference to the system of renting out serfs to mostly Jewish businessmen for three years at a time. Tens of thousands of Jews were massacred, with many more forced to relocate to the mostly barren territories to the east or to flee as far west as Alsace and Lorraine, leaving only a tiny Jewish population in Europe and worldwide Jewry at less than 1 million, its lowest point since biblical times.

Over the next century, the Jewish population gradually recovered, then began to grow exponentially. Most Ashkenazi Jews settled in the Ukraine or in Lithuania and Belarus, which were later annexed by imperial Russia. In 1791, Czar Catherine II ("the Great") carved out the infamous Pale of Settlement, the huge region extending from the pale, or demarcation line, of imperial Russia west to the border of central Europe—a continent-sized ghetto created to contain Jewish movement. Conditions were harsh; within the Pale, Jews paid double taxes and were forbidden to lease land, run taverns, or receive higher secular education. But that didn't retard the population boom. By 1900, the population of the Pale had grown to 5 million, and the Jewish population worldwide exceeded 10 million. The unique history of the Jews of Europe was now indelibly stamped in their genes.

ABRAHAM'S SONS

By the late 1990s, geneticists were finally in a position to unravel the mystery of the Jewish genome shaped by these periodic bottlenecks. Spurred by the Cohanim studies, an all-star roster of geneticists from the United States, Europe, South Africa, and Israel proposed drawing a definitive picture of Jewish ancestry back to the time of the diaspora that began in 586 BCE and even stretched back to the time of Moses. The scientists compared the DNA of nearly 1,400 Jewish and non-Jewish males from around the world—twenty-nine populations in all. Their findings, published in 2000, surprised many and delighted biblical literalists: Jews could trace their male lineage back to biblical Palestine. They share a distant, common ancestry with other ancient Middle Eastern populations, including Arabs and Palestinians, that predated the formation of Judaism. And despite having lived amongst gentiles for hundreds and in some cases thousands of years, Jewish males appeared to have mixed hardly at all with non-Jews after the founding of the Jewish population.

The massive study reinforced the Jewish historical belief that the core population of Jews has evolved as a biologically, as well as culturally, distinct family. "We really are a single ethnic group coming from the Middle East," University of Arizona geneticist Michael Hammer declared. "Even if you look like another European, with blue eyes and light skin, your genes are telling that you're from the Middle East."

The researchers found two primary Jewish Y-chromosomal ancestral lineages: the Med line, also known as the J haplotype, and YAP+4. The Med haplotype, which is where the CMH is found, is also common throughout the Mediterranean and in Europe and may have been spread by the Neolithic farmers or perhaps by the Sea Peoples. The slightly less common YAP+4 is believed to have originated in Ethiopia around twenty thousand years ago, before traveling along the Nile River and into the Levant.

The study concluded that Jews and non-Jewish Middle Eastern populations, particularly Syrians and Palestinians, were closely related, but more samples were needed to confirm this affinity. Each Arab and Jewish community did show a unique ancestral pattern, reflecting different genetic histories. Another common Jewish-Arab clade, known as E3b, is found among many Ethiopians; it may have originated in East Africa and then spread north along the Nile. However, today's Ethiopians, as well as the Falasha (the self-proclaimed Jews of Ethiopia, who are probably the descendants of postexilic converts), lack other markers that would distinguish them as ancestrally Jewish, based on their Y-chromosomal patterns.

Ashkenazi Jews were least like Russians, Austrians, and the British and most like Greeks and Turks. Jews from Muslim countries, Roman Jews, and European Jews had similar genetic profiles. It's thought the signature markers of the Iraqi, Moroccan, and Tunisian Jews may best represent the paternal gene pool of the ancient Israelites. The shared haplotype markers—the closely linked genetic markers present on the Y—among all the Jewish groups, except for the Ethiopian Jews, trace back four thousand years, which is about the estimated time when Abraham might have lived.

Do these findings cripple or confirm the Khazarian mass conversion theory? DNA rarely leads to such unequivocal conclusions. In fact, the study did find some Jewish groups with a haplotype variation that may have originated in central Asia. What could explain that?

Dror Rosengarten and Doron Behar, two of Karl Skorecki's protégés, set out to resolve the Khazarian mystery once and for all. Because of the bottleneck effect, even a few thousand converted Khazars should have left a noticeable imprint in the Ashkenazi genome. But the DNA picture is almost impossible to review, because there are no indisputable modern descendants of the Khazars. In hopes of finding genetic evidence in modern-day Eurasians, Rosengarten examined the DNA of men in two Jewish communities in the Caucasus—the Mountain Kavkaz Jews and the nearby Georgian Jews. No luck. The Mountain Jews matched the ancient Jewish profile, which indicates they emigrated from the Fertile Crescent, not from Khazaria. The Georgian Jews matched the genetic profile of non-Jewish Slavs and Europeans. They indeed may be a small population of descendants of converts, but they are not descendants of Khazars.

Behar then struck Khazarian gold, uncovering tantalizing genetic evidence in a study of Levites, who make up about 4 percent of the Jewish male population. In the second Cohanim study, in 1998, the researchers had found that while DNA has proved the Cohanim lineage originated in ancient times, there is no single common ancient Levite marker. Levites tested from around the world were found to have numerous lineages. If Jewish law had been maintained, they should share the markers with Cohanim, which is really just a branch of high priests among the larger community of Jewish Levite priests. For whatever reason, the ancestors of the Levites did not maintain a distinctive male line and oral tradition as rigorously as did the Cohanim. While close to 60 percent of those who claim to be Jewish priests can confirm they are part of an unbroken DNA line to a common ancestor thousands of years ago, only a few percent of Levites can. How could that be?

In his follow-up study, Behar found that the marker in more than 50 percent of Ashkenazi Levites does not even trace to the Middle East, is almost absent from Cohanim and Sephardic Jews (and Oriental Levites), and is found in only 4 percent of the general Jewish population. However, it is the dominant Y-chromosome haplogroup in eastern Europeans and nearby Eurasians, including regions once part of the Khazarian kingdom. His analysis of the fast-mutating microsatellites suggests the founding father dates to more than one thousand years ago—to central Asia, during the height of the Jewish Khazaria.

Intrigued, Ariella Oppenheim's Hebrew University team launched its own study and discovered the mutation in about 12 percent of Ashkenazim, which they speculated "may represent vestiges of the mysterious Khazars." They believe it traces back to 1,500 years ago to one or just a few Eurasian men. Analyzing the same data, Behar set the mutation's origin to approximately 650 years ago—coinciding with the final collapse of Khazaria and the migration into eastern Europe by the last remnants of the empire.

There is one additional piece to the Khazarian Jewish puzzle: what Behar has characterized as a "minor founding lineage" among the Ashkenazim, haplogroup Q. He found it in 5 percent of his subjects. Subsequent studies suggest it might not have its origins in the Middle East but in the general region of ancient Khazaria. It's also found in high percentages of Native Americans in Siberia and among Scandinavians, many of whom have ancestral roots in central Asia.

What do we make of these findings? What narrative could possibly conform with the genetic data?

In an attempt to answer that question—and this is only informed speculation—we return once again to the historical record. There is absolutely no credible evidence supporting the popular belief that Khazarians converted en masse to Judaism. We can assume that the Jewish Khazarian converts among the royalty and nobility might well have sought out the most distinguished status possible in the Jewish religious caste system—the honor of being a high priest, a Cohen. But that route to the priesthood is clearly blocked by Jewish law. However, restrictions on becoming a junior priest, a Levite, have historically been more flexible. After the Babylonian Exile, the Levites supposedly ignored Ezra's call to return as a group to Jerusalem, as the Cohanim did. They were summarily stripped of their exclusive rights to *maaser*, the tithe set aside for crops, which may have reduced the incentive to remain active Levites. Some scholars consider them fallen priests. Consequently, the paternal status of Levites was rarely enforced with the fervor exercised by the Cohanim. There are anecdotal examples of Jews claiming Levite status through the maternal line and even via conversion.

Could the rabbis who presided in Khazaria when the king and his court decided to convert have fudged the rules to bestow upon these new Jews the privileged status of the Levites? Considering the tiny size of the European Jewish population when Khazaria broke up, even this limited conversion, magnified by the bottleneck effect, could well explain the distinct and surprisingly significant imprint in the genomic record.

This scenario, however unprovable at this point, aligns with the genetic data. "If you go back to the Talmudic literature close to two thousand years ago, there's a whole discussion about other ways to enter the Levite status not through the father," said Skorecki. "There's no such route to the Cohen status. One can speculate that perhaps these other mechanisms might have been used in communities when there was a shortage of Levites, when there was an urgent need to have Levites, when something like that happened. Could that have happened in the case of the Khazars? It's certainly possible."

THE RUTH PHENOMENON?

What about Jewish women? The belief that most Jews are descended from the ancient Israelites has turned out to be a very male-centric perspective. While most of the founding fathers of the Jewish people trace their ancestry to the Middle East, the origin and history of the founding mothers is more of a mystery, only slowly being unraveled. The DNA evidence suggests that many Jews along their maternal line may be descendants of converts.

If that theory holds, it shouldn't be surprising. Throughout history, women have often assumed the religion of their husbands, and that's especially true in the Bible. Jacob's son Judah married a Canaanite (Genesis 34:9); his brother Joseph married an Egyptian, Asenath (Genesis 41:45) and had two sons by her, Menasseh and Ephraim, who would become patriarchs of two of the Twelve Tribes; and Moses married Zipporah, the daughter of a Midianite priest (Exodus 2:21), and then possibly a Cushite (Numbers 12:1).

One of the most eloquent biblical expressions of what it means to be Jewish comes not from an Israelite but from a gentile—Ruth. The book of Ruth has long been considered one of the most stirring and tender tales in the Bible and an inspiration for converts to Judaism. In the days of the judges, before the royal dynasty of Israel was established under Saul, the tragic story of Elimelech and Naomi of Bethlehem unfolds. To escape the famine that has devastated their homeland, the patriarch resettles his family in the land of the Moabites, bitter rivals of the Israelites. He soon dies, after which his two sons marry local women, Orpah and Ruth. When the sons also die unexpectedly ten years later, after the famine has spread to Moab, their mother, Naomi, and her two daughters-in-law are devastated and destitute. Naomi decides to return to Israel, where the barley harvest is commencing, and beseeches Orpah and Ruth to remain with their people.

The story takes a moving turn, as Ruth and Naomi establish an affectionate bond and a biblical model of friendship and devotion. While Orpah stays behind, Ruth pledges her loyalty to Naomi and declares her eagerness to embrace the alien faith of the Israelites. "Wherever you go, I will go," she says in Ruth 1:16. "*Ameikh ami, ve'Elo-hai-ikh Elo-hai*—Your people shall be my people, and your God my God." With Naomi's encouragement, Ruth marries her mother-in-law's cousin, the prosperous Boaz, who would become the great-grandfather of King David.

"The Book of Ruth has long served as an important antidote for any Jew prone to exaggeratedly nationalistic leanings," Rabbi Joseph Telushkin has said. "How chauvinist can one become in a religion that traces its Messiah to a non-Jewish convert to Judaism?"

That Israelites mixed with gentile women should hardly shock those familiar with patriarchal cultures stretching back thousands of years. The descendants of Abraham were never "pure" along the female line. In early biblical times, a woman was a form of property, and marriage was an acquisition, with neither God nor the state having much say in the matter. Although the Bible says nothing about the marriage customs of the Israelites during their centuries in Egypt, passages suggest marriage outside the clan was common. Leviticus 18 and 20 discuss a number of sexual taboos, but the mixing of gentile and Jew is not one of them. The Hebrews consisted of an amalgam of Semitic and maybe other lineages, and polygamy was rampant. Jewish men often "dwelt among" other tribes, taking local wives.

The question of which tribal clans were acceptable for providing wives and which were not varied according to the circumstances. Although Abraham did not want his son Isaac to marry a Canaanite, it was not a fundamental reaction against intermarriage. Rather, he was greatly concerned that marrying a Canaanite would sever the Lord's commitment of passing along the priesthood. Centuries later, when Moses left Egypt with his flock, he brought along a "mixed multitude" of tribes, including many who were not Jacob's descendants. When the Hebrews defeated the Midianites in battle, Numbers 31:31–34 relates how Moses at God's direction allowed them to keep as "booty" all "the women who had not had carnal relations," 32,000 virgins in total. Even Deuteronomy, which frowns on the taking of gentile wives, notes that God allowed a warrior victorious in battle to "possess" any "beautiful woman" who should catch his fancy as long as she follows certain rituals and restrictions (Deuteronomy 21:10–13). This debate over the racial purity of the early Hebrews seems a bit ridiculous, as there was an endless supply of gentile virgins around for the picking.

According to the Bible, this tribal mixing continued after the return of the Hebrews to the Promised Land. "The Israelites settled among the Canaanites, Hittites, Amorites, Perizzites, Hivites, and Jebusites," reads Joshua 3:5–6. "[T]hey took their daughters to wife and gave their own daughters to their sons, and they worshipped their gods." The taking of Moabite wives, like Ruth, was hardly unusual. King Solomon, supposedly the wisest and perhaps the most polygamous Israelite ever, "loved many foreign women," including "700 royal wives and 300 concubines," according to 1 Kings 11:1–3. Rather than convert them, Solomon is said to have built them idolatrous temples. Such practices eventually brought down the wrath of God (and ironically provided a rationale in Jewish lore for switching the measure of Israelite identity from the Y chromosome to the X).

Sometime between 200 BCE and 500 CE, the uniquely Jewish reliance on the womb to define religious identity emerged. Although some contemporary historians maintain that tying Jewishness to the birth mother has its genesis in Ezra, the Harvard professor Shaye Cohen maintains that the rabbinic edict originates in Roman traditions absorbed by the Jews. According to Roman law, later mimicked in the Mishnah, intermarriages between citizens and noncitizens produce children whose status is determined matrilineally. It was only with the dramatic spread of Islam that most Jewish communities adopted the Talmudic teaching of the matrilineal definition of Jewishness.

If the Christian gospels are fair measure, the matrilineal principle may have been in place in Palestine as early as the end of the first century CE, when Christianity was still mostly a Jewish sect. Consider the different ways that Paul treated his two young preachers, Timothy and Titus. Timothy was the son of a Jewish mother and a gentile father. Paul circumcised him because he was Jewish. Titus had a Jewish father and a gentile mother and was considered non-Jewish. Paul denied him circumcision and denounced those who felt it necessary for him to have the rite.

The Ruth theory rests largely on the 2002 Center for Genetic Anthropology study of nine geographically separated Jewish groups. It found that most Ashkenazi Jewish women appear to be descended from non-Jewish Europeans. There were no signature mutations found on the female DNA—no significant founding event—as there were for Jewish men, and few recent genetic ties to the Middle East along the female line. David Goldstein, a key participant in the study, likened central and eastern European Jewry to a genetic mosaic of separate, small populations. "The men came from the Near East, perhaps as traders," he speculated. It appears that Jewish men often didn't bring along spouses. "They established local populations, probably with local women. But once the community was founded, the barriers had to go up, because otherwise mitochondrial diversity would [have shown in the study to be] increased." Each of the Jewish villages appears to have been independent of one another, and once formed, slammed its doors shut to new converts. If this explanation holds, based on the matrilineal determination of Jewishness, many of the founding mothers of Ashkenazi Jews may be the descendants of righteous converts or maybe even women living as Jews who never converted—by Israeli law, not Jews at all.

There's been a friendly war over those results between the London team and the Haifa researchers. Doron Behar was convinced the results of the London study might have been skewed because the study didn't sample enough women. The Technion study analyzed 1,700 women. Behar found evidence of a founding event that roughly coincides with the beginnings, during the first millennium, of the Ashkenazi Jewish community in Europe, probably in the Rhine basin. He also found that about 4 million of today's Ashkenazi Jews—about 40 percent of Jews originating in central and eastern Europe—descend from just four women. A series of Jewish population expansions and contractions in Europe over many hundreds of years beginning in the thirteenth century spread the marker mutations throughout various Jewish communities.

What is not clear are the original origins of these Ashkenazi Jewish females. They could trace to Israelites who migrated to Europe. There is also some evidence of mixing between Ashkenazi Jews and Jews expelled from Spain who relocated in Europe. Or, in concert with the British findings, the marker mutations could be from European or Eurasian women—supporting the local wives theory. Behar believes that most of these four lineages representing 40 percent of European Jewish females probably originated in ancient Mesopotamia. The other 60 percent of Ashkenazi Jewish women more than likely have European and some Middle Eastern roots.

These findings do comport with biblical history. "This is very consistent with the tradition that the early Israelites took wives from a wide geographic range," said Karl Skorecki. "Certainly, the mtDNA of Sara, Rivka, Rachel, and Leah (and Bilha and Zilpa) were not expected to have been passed on to the descendants of Jacob exclusively. This is evidence consistent with the matrilineal definition of Jewish identity in practice for a very long time, throughout the world."

Taken together, the Jewish male and female lineages offer a fascinating, if controversial, narrative of the ancestry of Ashkenazi Jewry, even though it is based on only two loci, the male and female genetic marker—very tiny, if revealing, slices of the human genome. The studies of the Y chromosome and mtDNA do not support the once-popular notion that Jews are descended in any great numbers from the Khazars or some Slavic group, although it's evident some

Jews do have Khazarian blood. The Khazarian theory has been put to rest, or at least into perspective. Perhaps not every Jew is descended solely from the ancient populations in Judea and Samaria, as the Bible suggests, but most Jews do share a common ancient ancestry. Most Jewish males appear to have originated in the eastern Mediterranean, with at most 20 percent showing a central Asian origin similar to that of most Europeans. After being expelled from the Middle East, and after diaspora stops along separate routes in Italy and Asia, Jews trickled into Europe. They brought with them some wives, but more often than not, they coupled with local women.

Assuming eighty to one hundred generations have passed since the founding of the Ashkenazi population, Michael Hammer has calculated that after the initial trysts and founding of various Jewish villages, less than 0.5 percent of each succeeding generation of Ashkenazi women had children with non-Jewish Europeans. The initial conversions and the modest intermixing over many generations has had an impact on adaptive characteristics like blood groups, facial characteristics, and skin color, which helps explain why Jews often tend to physically resemble non-Jews in their host communities. (How many times have you heard the phrase "But you don't look Jewish!")

Jews, like only a few other culturally distinct populations—Mormons and the Amish come to mind—have certainly stuck to their knitting, at least until recent decades. "The fact that we don't see [signals of genetic mixture between Jews and non-Jews] suggests that after the diaspora these populations really have managed to maintain their Jewish heritage," said the University of Leicester geneticist Mark Jobling. Oppenheim and her colleagues agreed, writing that they had "expected a few more admixtures." Oppenheim had anticipated seeing in the data the bloodlines from the rape of Jewish women that occurred during pogroms in eastern Europe. "It had an effect," she wrote, "but it did not significantly alter the gene pool."

Jews have received these studies confirming their relative purity with a combination of enthusiasm and concern. They reaffirm the central threads of Jewish history and identity: diaspora Jews are ancestrally connected to each other and to biblical Israel, a belief that binds religious, nonpracticing, and even atheist Jews. But the data are not without yellow flags. "This is exciting research," says Lawrence Schiffman, the chairman of the Department of Hebrew and Judaic Studies at New York University and an Orthodox Jew. "It demonstrates that the cohesive spiritual and educational heritage of the Jewish people has been maintained throughout the ages. It will certainly play a role in how Jews define themselves. But it's going to challenge some long-held academic and personal religious views," he added. "We are in this wrestling match in Judaism about to what extent we are a religion or a people. It's going to make some people nervous. Whenever the words 'Jew' and 'race' are used together, even with all the scientific qualifications, there's room for possible trouble."

PART III

RACE

CHAPTER 10

JEWS AND RACE

For much of their history, Jews have been bedeviled by the notion of a "Jewish race." Most Jews are proud of their tribal history and cultural distinctiveness, and it's been a key factor in the survival of the Jewish people. But it's also been a source of friction and worse to non-Jews, who have often viewed Jewish chosenness and exceptionality as cultural chauvinism.

Over more than two thousand years, in each generation, in each far-flung Jewish community, Jews have had to calibrate, often contentiously amongst themselves, the benefits and costs of separation and assimilation, of living by the Book or in the world. These internal struggles for the soul of Judaism have led to fractious battles within the Jewish community and between Jews and other groups—among the northern and southern kingdoms; the Judeans and the Samaritans; the secular-minded Hellenized Jews and the traditionalists; and the Jews and the Jewish Christians, who would break away to create a new religion and who would set aside the notion of a God whose truths were embedded in sometimes inscrutable texts.

The fate of Jews in the Middle Ages would once again test that historical tension in Judaism between religiosity and reform. During the Protestant Reformation, as Enlightenment reforms began leavening Christian orthodoxy, the ultrareligious Ashkenazi Jews appeared to Christians as hopelessly backward. The state of world Jewry was fragmented and leaderless. Many Jews dressed differently and kept to themselves. The Judaism they practiced was focused inward and torn between strict adherents of the Talmud interpreted by powerful rabbis and Jews seduced by kabbalism, the mystical form of Judaism that stressed a nonrational religious experience often tinged with demonology and messianism. Half or more of all Jews, commonly the poorest, embraced the charismatic ravings of the Ottoman Jew Shabbetai Izevi, who claimed that he was the Messiah. Kabbalism later became formally known as Hasidism (from the Hebrew for "those who are pious"). Although it began as a challenge to the intellectualism of Talmud study, it would evolve into a haven for traditionalists opposed to religious modernization—a place Hasidism still holds in Judaism.

The image of the slightly depraved Ashkenazi Jew persisted in the imagination of Christians for some time. For François-Marie Arouet, better known as Voltaire, the great French philosopher and champion of reason, the Jews were a race—and a hopelessly unsalvageable one. "We find in them only an ignorant and barbarous people, who have long united the most sordid avarice with the most detestable superstition and the most invincible hatred of every people by whom they are tolerated and enriched," he wrote in "Juifs," an article from 1762 that was later included in his *Dictionnaire philosophique*. Many entries in this dictionary derided Judaism, which he viewed, not without some justification, as a fossilized metaphor of tribalism, scripturalism, legalism, and insularity.

Even defenders of the Jews, such as the French priest and parliamentarian Henri Baptiste Grégoire, who advocated abolishing Negro slavery and granting citizenship to Jews during the constitutional debates in the 1790s, speculated that Jews possessed a pathological uniqueness that might explain their ethnic longevity. Grégoire later wrote that Jews were "naturally gloomy and melancholy" and were particularly libidinous, especially the women, who were obsessed with sex because of the "accumulat[ion of] many acrimonious particles in the mass of humors contained in their bodies." As for whether Jews could ever become part of mainstream Europe, Grégoire was not optimistic. "Climate has scarcely any effect on them, because their manner of life counteracts and weakens its influence," he wrote. "Difference of periods and country has, therefore, often strengthened their character instead of altering its original traits." French Enlightenment thinkers were joined by German idealists, from Immanuel Kant to Georg Wilhelm Friedrich Hegel, in their disdain of Ashkenazi Jewry's obsession with biblical ritual.

SEPHARDIM V. ASHKENAZIM

The cultural stereotype of the downtrodden, religiously atavistic Ashkenazi Jew stood in marked contrast to the reputation of Sephardic Jews, who had been forced out of Spain and Portugal during the Inquisition. Shortly after the publication of "Juifs," a Dutch banker, Isaac de Pinto, challenged Voltaire's depiction of Jews. What offended him was not so much Voltaire's low opinion, which de Pinto shared, as his failure to draw a distinction between what he believed were two separate "races" of Jew: Sephardim and Ashkenazim. It's no wonder de Pinto was so upset; Voltaire's famous disputant was in fact a Spanish crypto-Jew, one of hundreds of thousands of Sephardim whose families had been forcibly converted to Christianity but had secretly kept alive their Judaism.

"A Sephardic Portuguese Jew from Bordeaux and an Ashkenazi German Jew from Metz appear to be two entirely different beings," de Pinto wrote. He suggested that Sephardim were characterized by an "elevation of mind" and retained cultured qualities that were altogether compatible with Christian Europe and which distinguished them from the Ashkenazi Jews, many of whom were desperately poor and backward.

Until the Inquisition and the massacres, conversions, and intermarriages that accompanied it, the large population of Iberian

Jews had long been thought of, and thought of themselves, as culturally and even biologically superior to eastern European Jews. "The Sephardic looked down from on high upon the poor little Jew from the north," wrote Joseph Nehama in *Histoire des Israélites de Salonique*, his study of this historical period. In contrast to the educated Sephardim, it was the Ashkenazi "accustomed to misery and oppression, who cowered, made himself humble and hugged the wall, who had always lived with doors and windows shut, who shunned all social contact, all friendship with the non-Jew, who, always despised and unwanted, lived here and there as a veritable nomad, always ready to take off, with his bundle and his wanderer's staff."

The movement of exiled educated Sephardim, New Christians, and *conversos* into the Low Countries and sections of western Europe that were relaxing restrictions on accepting Jews gradually changed the perception of Jews. What can only be called an era of "philosemitism" sprang up in the late seventeenth and into the eighteenth century in the salons of Antwerp, Amsterdam, London, Hamburg, Frankfurt, and Bordeaux. While Jews had long been viewed warily for their moneylending practices, educated Jewish merchants and bankers were fancied as models of thrift and self-sufficiency—at least until anti-Jewish sentiment later resurfaced.

Amsterdam become known as "the Dutch Jerusalem"—an international center of commerce and humanism. Sephardic artisans helped found the diamond trade, and exiled Jewish seafaring merchants played an active role in the Dutch East and West India companies, which helped settle New Amsterdam—New York City. As a symbol of their integration into Dutch society, the large Portuguese expatriate community would eventually build a magnificent new synagogue, inaugurated in 1675. This interim softening of attitudes was accompanied by a surge of interest in the Lost Tribe myths, which spurred curiosity about biblical and postbiblical Jewish history among Christian millenarians.

When welcomed, Sephardic Jews relaxed their separatist ways. This tolerance left its mark in the human genome. Studies show that Dutch Jews are the most genetically mixed along the male line of any Jewish group in Europe, a testament to their easy relationship with Christians as well as evidence of the mixing of the often-separate Ashkenazi and Sephardic communities. Along the female line, Dutch Jews do not appear to have increased levels of European female DNA, suggesting that most of the intermarriage was between Jewish women and non-Jewish men—the opposite of what geneticists have found occurred in Ashkenazi Europe.

Baruch Spinoza tried desperately to take advantage of this era of tolerance to modernize Judaism. Born in Amsterdam in 1632 of Spanish-Portuguese parents, he offered a decidedly Jewish interpretation of the humanist and scientific writings of the early Enlightenment Christian thinkers. A carrier of the flame of Moses Maimonides, Spinoza saw no distinction between God and Nature, basing his views on the dramatic scientific discoveries of Isaac Newton and the philosophic writings of Thomas Hobbes in England and René Descartes in France. In his later writings, he challenged the dogmatic interpretations of Jewish texts then in vogue and laid the foundations for biblical criticism by asserting that the Torah was not the revealed Word of God, but a human document sprinkled with stories of apocryphal miracles that were inserted by scribes many centuries after the time of Moses. Spinoza's sharp critique of the pretensions of Scripture and sectarian religion broke sharply from the medieval supernaturalism that gripped Ashkenazi Jewry and Christianity.

However, even in the progressive Sephardic Jewish community, Spinoza's deism—the belief that the Creator had removed Himself from involvement in the material world and human affairs—branded him as dangerous. His views were initially not well received by either Jews or skeptical Christians. While in his mid-twenties, Spinoza was banished by the rabbis of Amsterdam. His excommunication—known in Hebrew as *kherem*—was absolute. He was not allowed to do business with, talk to, or even stand next to a Jew. Considering that Spinoza's critique became entirely mainstream among the educated within a century, the stain on his reputation has always rankled liberal Jews, often to no avail. (When the Israeli prime minister David Ben-Gurion suggested in the 1950s that the rabbis of Amsterdam rescind the *kherem* against Spinoza, he was rebuffed.) However heretical at the time, Spinoza's views would eventually take root and reform the Jewish religious establishment that turned on him so viciously.

THE JEWISH ENLIGHTENMENT

By the late seventeenth century, Christian Europe had experienced a recovery of nerve, shaking off centuries of barbarism and intolerance to embrace the belief that different cultures or religions could have similar merit. But it would take many more decades before the Enlightenment principles would open the way for the renaissance of Ashkenazi Judaism. In the 1740s, Jews in the Prussian-controlled region of Galicia became the first large Jewish community to adopt many of the secular reforms that had already stormed through Christian Europe. The movement, known as the Haskalah, the Jewish Enlightenment—it comes from the Hebrew word *sekhel*, meaning "reason" or "intellect"—would eventually spread throughout central Europe. After centuries as specks on the map of Europe, the settlements of those practicing the transformed Judaism were poised for explosive growth.

While Spinoza embodied the early Enlightenment efforts at modernizing Orthodox Judaism, the German Talmudic scholar Moses Mendelssohn symbolized the fruits of that effort. He was born in 1729. In authoring the first translation of the Bible into German, he helped dramatically upgrade Ashkenazi Jewish educational standards. Highly educated, comfortable around Christians, and a great supporter of the German language and culture, Mendelssohn presented a personal and direct challenge to the prevailing stereotype that all Ashkenazi Jews were superstitious, secretive, and separatist. He became what might be called the "house Jew" of

Christian intellectuals, a favorite example of how they wished all Jews might act.

The Jews of greater Germany, eager to become a part of the secular life now blossoming around them, embraced Mendelssohn's goal of integrating themselves into Enlightenment Europe. Reformers challenged the medieval state of Jewish education still mandated at most yeshivot, reviving biblical Hebrew as an antidote to ghetto Yiddish and stressing science over Talmudic debates stretching back to biblical times, such as how and on what days to gore oxen, which had no relevance to modern Jews. The emphasis of Judaism began to shift to match the Christian model—Jews lived their daily lives based on reason and looked to faith rather than chosenness as the moral basis of their religion.

While the Enlightenment would pass the mantle of Jewish culture from Spain to central and western Europe, it was slow to transform the ghettoes that had defined Jewish life in eastern Europe for centuries. The pace of change for the disenfranchised Jews of Russia and Poland confined to living mostly in village shtetls proceeded extremely slowly. Hardened by years of persecution, eastern European Jews resisted being absorbed by their new host states. By 1791, Czarina Catherine II had abandoned her efforts toward integration, carving the Pale of Settlement (encompassing parts of present-day Latvia, Lithuania, Ukraine, and Belarus) out of the eastern flank of the carcass of Poland. Many Orthodox rabbis staunchly opposed reforms because they challenged both rabbinic authority and the commanding role of the Talmud in Jewish life. This steadfastness fed Jewish stereotypes. Writing in 1898, the German Jewish scholar Heinrich Graetz noted that Polish Jews display "a love of twisting, distorting, ingenious quibbling" that he believed resulted from their "cultivation of a single faculty, that of hair-splitting judgment [from studying the Talmud], at the cost of the rest, narrow[ing] the imagination, [without] a single literary product appear[ing] . . . deserving the name of poetry."

Although the Jewish communities of the Pale and western

Europe were moving in opposite directions, when it came to religious orthodoxy, they shared a broad commitment to education. "Almost every one of their families hires a tutor to teach its children," wrote an Orthodox Christian Russian official in 1818 about the villages he oversaw. "Their entire population studies. Girls too can read, even the girls of the poorest families. Every family, be it in the most modest circumstances, buys books, because there will be at least ten books in every household," he marveled. The official ruefully contrasted the standards among even the poorest Jews to the low literacy rate in Christian communities. "Most of those inhabiting the huts in [gentile] villages have only recently heard of an alphabet book," he wrote.

Owning books was extraordinary because they were so expensive and had little utility for most people, who lived on farms. But many Jews put aside their economic self-interest to follow the dictates of their religion, which imposed the unique obligation to read daily from their holy texts.

Based on the surviving genealogies of this period, the families of rabbis and merchants tended to have many children—eight or nine in some cases—while poorer families understandably had fewer children able to survive the rigors and epidemics of the ghetto. Geneticists would later cite this as a form of positive selection that might explain, in part, the outsized record of Jewish intelligence and professional accomplishment—the best and the brightest of Jews passed on their genes, while the poorer and less educated had fewer or no children.

Rising expectations and the withering away of feudalism inspired two of the most pivotal events to shape modern Jewish history—the French Revolution and the founding of the United States. America was the first country to grant Jews full and equal rights. France, an eighteenth-century hotbed of republican ideals and a key supporter of the American insurgency against Britain, needed a revolution of its own before the barriers to reform would crumble. When the Bastille, the grim fortress that symbolized the ancien régime, fell in 1789, there were approximately forty thousand Jews in France. Although they embraced the republican triumph in Paris and pledged their fidelity to France, many Christians continued to view them with suspicion. They were seen as more loyal to their community and "their God" than to the French flag and the new cause of universal brotherhood. Jews were a "nation within a nation," in the words of Stanislas, comte de Clermont-Tonnerre, the president of the French National Assembly in 1789.

After a contentious debate over granting Jews more rights, it was decided that Jews would be welcomed as citizens of the "new France." "The Jews should be denied everything as a nation but granted everything as individuals," Clermont-Tonnerre declared. Special taxes were eliminated, and the ghettoes were ordered liquidated. Although heartened, most Jews did not immediately step out from the safety of the wall-less ghetto. Hardened by centuries of persecution and decades of false promises, they were wary of dropping "peoplehood" as a defining component of Judaism. Frustrated, Napoléon in 1806 presented Jews with a historic choice: if they would accept what he called their "full" responsibilities as Frenchmen and switch their primary loyalty from ancestry to nation, he would tear down the legal barriers to full citizenship. Secular and religious convocations of French Jews agreed overwhelmingly to establish the primacy of French law and refused to budge only on the question of sanctioning intermarriage. Jews were henceforth to be seen not as a nation, but as "French citizens of the Mosaic faith." The intermarriage issue aside, the Jews of France began thinking of themselves as French.

Jewish legal emancipation swept through nineteenth-century Europe, although it took different forms in each country. In 1812, the Prussians accorded the Jews many civil rights, although they restricted them from government service. Denmark opened the doors for Jews, in 1814, to enter any profession. Italy tore down its ghetto walls in 1848, the same year that the German National Assembly proclaimed the full rights of all of the Jews. The following year, the Austro-Hungarian Empire passed a new constitution that extended equal rights to Jews (although it was suspended between 1851 and 1867). Bavaria became one of the last regions to drop restrictions, which effectively completed the legal emancipation of European Jewry.

The Haskalah was transforming the fossil of Ashkenazi Jewry, but the revolution was not without peril. Rabbinical Jews warned that the glue that held this ancient people together in exile for so many centuries—the central belief in ancestry and divine chosenness—was in danger of being swept away by the forces of secularism and intermarriage. And they were right. Hamburg, which had been a magnet for educated Jewish tradesman of Sephardic ancestry, became the center of a movement known as Reform Judaism. Synagogues began introducing sermons, choirs, and organ music, while rabbinical authorities revised the prayer books and ended the forced separation of men and women congregants. Reformers coined the term "ethical monotheism" to underscore the social responsibility of Judaism and Jews and the rejection of the ritualism of the Talmud.

It was only a matter of time before emancipation led many Jews to question their faith, loosening their cultural and religious ties. Within a few generations, most German Jews had stopped keeping kosher, and the prohibition on mixed marriages had faded. In a sign of the times, the Reform congregation of Berlin even shifted the Sabbath service to Sunday to synchronize with the services of their Christian neighbors. Although the Jewish birthrate in Germany improved in the nineteenth century, the percentage of Jews in the population actually declined, as many stopped practicing or converted. Many practicing Jews were more apt to identify with their German nationalism than their Jewishness.

In 1840, while world Jewry was catalyzed by the plight of the Jews of Damascus, who were accused of a trumped-up blood libel, one of the first ever in Muslim lands, German Jews were more intent on affirming their national bona fides. "For me, it is more important that Jews be able to work in Prussia as pharmacists or lawyers than that the entire Jewish population of Asia and Africa be saved," said Abraham Geiger, who would soon become Germany's chief Reform rabbi. The intelligentsia, which referred to itself as "Israelites" at the turn of the nineteenth century, was now going by "German Citizens of the Mosaic Faith."

A wave of economic and political instability that rolled through midcentury Europe accelerated the secular reforms. A series of potato and wheat harvest failures sent food prices spiraling. Cholera broke out. The crisis escalated after an international credit collapse, which led to wholesale bankruptcies. This perfect storm destabilized governments across Europe, many of which fell to revolution. The troubled times spawned a generation of radical Jewish intellectuals and activists, with Karl Marx the prototypical example. Born in 1818, Marx was descended from a long line of conservative Jewish rabbis. But secularism would fracture the family. While his uncle became chief rabbi in Trier, Karl's father, Herschel, converted his family to the Prussian state religion of Lutheranism to advance his legal career. Karl never had much time for Judaism or any religion. He considered religion, in his famous phrase, as "the opium of the people" and blamed middle-class society for buying into "bourgeois capitalism," which most liberal European Jews, long ostracized from mainstream society, were yearning to be a part of. Marx's revolutionary formulations would become a cultural signature of European and American Jewry for many years.

The economic dislocations prompted many Germans, Jews included, onto transatlantic ships ferrying Europeans to a new start in the capitalist promised land of America. Bohemian-born Isaac Mayer Wise arrived in 1846, one of hundreds of thousands of German Jews who relocated to America between 1830 and 1870. After a brief time in Albany, he settled in Cincinnati, where he founded the Hebrew Union, still a scholarly college and center of Reform Judaism. The Reform movement became the backbone of liberal American Judaism. Rather than being schooled in Jewish yeshivot, German American Jews enrolled in the free public school system, ensuring their mixing in the American melting pot. Would this experiment in assimilation complete the transformation of Judaism from a religion of ancestry into one of faith?

FOLK RACISM

By the nineteenth century, for the first time in their history, educated Jews dared to think of themselves not just as a people waiting to return to their ancient homeland but also as active participants in civil society. They left their rural villages for the big city to embark on once-unheard-of careers as lawyers, doctors, and even teachers. Increasingly confident, they became an integral part of the bourgeoisie. Influential Jewish communities flourished in Paris, Vienna, Prague, and Berlin. The merchant noblesse of Europe were now dotted with Jewish tycoons—the Oppenheimers and Wertheimers of Austria, the Pereiras in France, the de Hirschs of Bavaria, and the Bleichroeders in Prussia; the moneylending business started by the Rothschild family in Frankfurt flowered into a Europe-wide financial empire.

In this era of rising nationalism, increasing numbers of Jews assimilated into their regional cultures, but few totally lost their Jewishnesss. Even Jews who vigorously attempted to discard their Jewish identity, including Sigmund Freud, remained doggedly Jewish in many stereotypical ways. Like many Austrian Jews of his time, he made a very public demonstration of his pan-German commitment, joining the ultranationalist student society Leserverein der deutschen Studentum. He was an avowed atheist, but he was educated about his Jewish heritage, had only a few friends who were non-Jews, and regularly attended meetings at his local *Gemeinde*, the officially sanctioned Jewish bodies that oversaw communal affairs.

The German-speaking countries emerged as the wellspring of Jewish achievement, particularly in the sciences. Jews were visibly represented in the liberal professions. Berlin and Vienna became the intellectual hothouses of Europe. Berlin was home to the Prussian Academy of Sciences and later the Kaiser Wilhelm Institute, which attracted the best and brightest of European scientists, an extraordinary number of them Jewish. By the 1880s, almost 25 percent of Viennese law students and more than 40 percent of the medical students were Jewish. By the turn of the century, Jews totaled 50 percent of the doctors in Berlin and 60 percent in Vienna.

The enormous success of Jewish assimilation stirred a powerful backlash. Just as Jews were breaking taboos by demonstrating that they could contribute to secular Christian society, deepening social cleavages called forth all sorts of conjecture about class and race. A low fever of anti-Jewish prejudice began heating up in Germany after the founding of the Christian Social Party in 1878. Laying the groundwork for Nazi National Socialism, its leader, Adolf Stöcker, a former army chaplain, advocated a hodgepodge of economic reforms and wrapped them in virulent, ultranationalist rhetoric. He blamed "foreigners," among whom he counted Jews, for the economic unrest plaguing Europe. Although they were succeeding in "mixed society" and were proud of their German heritage, educated Jews commonly maintained a strong ethnic pride and affiliations with local Jewish institutions and work guilds, which bedeviled many Germans.

The assassination of the reform-minded Russian czar Alexander II in 1881 only exacerbated tensions in eastern Europe. Bloody pogroms sent millions of peasant Ukrainian Jews streaming westward or onto boats bound for the Americas. The lumpen eastern European Jews arriving in Hungary, Austria, and Germany mixed like oil and water with the established class of acculturated Jews. The Dark Ages met the Enlightenment, and the picture was not pretty in the eyes of European Christians. By 1910, Vienna saw its Jewish community balloon to more than 180,000, almost 10 percent of the population—many of them eastern Europeans who looked to the locals like misfits. It was not long before the presence of these oddly dressed Jews with beards and sidecurls prompted Europeans to talk openly about *der Judenfrage*—the "Jewish question"—the Jewish penchant for separatism, which Christians believed persisted even when Jews were faced with little persecution. Many Christians, few of whom considered themselves anti-Jewish, claimed they would have been more tolerant if Jews had not continuously asserted their exceptionality and their right to remain a theocracy within a state.

Anti-Jewish paranoia was whipped to a frenzy by the widely circulated anti-Jewish tract *The Protocols of the Learned Elders of Zion*, which supposedly documents a cabal of rabbis and Jewish secular leaders plotting in secret to take over the Christian world. The seeds of the infamous fabrication probably originated with the blood libels of the Middle Ages, and it then was updated by hired anti-Jewish reactionaries in Paris during the height of the hysteria that accompanied the Dreyfus Affair. It was later circulated by the czarist secret police, which found it useful in its efforts to discredit liberal reformers, who were gaining popular support among oppressed minorities, such as Russian Jews. As Christopher Hitchens has written, the premise that it was authored by Jews plays like a sick joke. "The Jews are supposed to be diabolical and clever enough to plot a secret world rule, and stupid enough to write the whole plan down," he noted wryly. "The Jews' plan is that, from being the most despised and reviled minority in history, they go straight to a worldwide takeover and supreme power. Just like that."

Germany and to a lesser degree the rest of Europe and North America were in the grip of folk racism. Medicine, nationalism, and the pseudoscientific leanings of anthropology had become hopelessly entangled. The "Jewish question" became entangled in the debate over eugenics, a social philosophy that humanity could selectively improve its physical, moral, and intellectual health through selective breeding. It was first, crudely, articulated in 1851 by Herbert Spencer, then an editor at the *Economist* in London. At the time, Europe was a world of the haves and have-nots, with the poor locked in squalor. Spencer saw material success as a sure sign of superior stock. The plight of the downtrodden was nature's way of selecting the hardiest to breed an even better race of humans. Eight years later, Charles Darwin, in *The Origin of Species*, settled on the less grandiose thesis that natural selection drove living beings not to thrive but merely to survive.

Their complementary theories excited the imagination of Darwin's cousin, Francis Galton, who in 1865 coined the term "eugenics," from the Greek words meaning "good birth." In his most important book, *Hereditary Genius*, published in 1869, he juiced up natural selection, loosely applying its principles to the mind in an attempt to ground psychology in science rather than speculation. He claimed that intelligence and ability were products of the same evolutionary laws that shape the natural world and that Europe's colonial success reflected evolutionary forces in the same way nature determined that lions ruled the jungle. Blacks were considered "childish, stupid, and simpleton-like" while Jews were merely "crafty," a far cry from the genius of elite European white males. The fashionable discipline of what became known as social Darwinism appeared to explain the underlying cause of poverty and other human social problems, delivering as a corollary a sobering belief that class and racial conflicts were inevitable.

This mystical fusion of nationalism and racial chauvinism that pervaded Victorian Europe would come to haunt world politics. Political extremists, from Marxist communists on the left to ultranationalists on the right, would exaggerate social Darwinism and its offspring, eugenics, to justify a romance with racial politics. Even Jews were not immune. In Victorian Prague, Vienna, Berlin, Paris, London, and even New York, Jews were no less insistent than Christians that they constituted a biologically distinct people. However, while many gentile scientists insisted that Jews were degenerate and lower on the ladder of civilization than European whites, Jewish scientists claimed racial purity as a way to restore the tarnished dignity of the "people of the Mosaic faith" under attack by Christian racialists.

The Australian-born British anthropologist Joseph Jacobs was the first outspoken proponent of reviving Jewish biological exceptionalism. In his 1885 paper "The Comparative Distribution of Jewish Ability," he argued that the ghetto had worked as a kind of genetic jungle, where only the fittest survived. By this theory, the often-lachrymose course of Jewish history was endowed with a redeeming existential purpose. Just at the time Jews were under increasing suspicion as "foreign elements," the Jewish elite was insisting on reconnecting Judaism with the historical sense of peoplehood that had long been a source of their tensions with non-Jews.

Although eugenics was first articulated in Britain, it found its most ardent converts in Germany, which embraced selective breeding as a kind of racial Holy Grail and directed it most pointedly at Jews. In 1905, Professor Alfred Ploetz launched the German Gesellschaft für Rassenhygiene, the Society of Racial Hygiene. Although they constituted only 1 percent of the population, Jews were Germany's largest minority. Ploetz, a leading socialist of the time, was convinced that tensions he believed Jews provoked could be resolved if Jews would only allow themselves to be absorbed into German society and not insist on retaining their separateness. He and his followers did not consider themselves anti-Jewish, were generally not considered so by Jews, and in fact trained Jewish physicians with hopes of encouraging the spread of assimilationist thinking.

One of the first prominent German Jews to take up the cause of eugenics was Elias Auerbach, a physician from Berlin. Writing in one of the prestigious academic eugenic publications of the day, in 1907, he addressed what he believed was a distressing trend toward intermarriage, which was watering down the "Jewish race." By his estimation, intermarriages then accounted for one-sixth of all unions involving German Jews. He traced Jewish biological distinctiveness to Jewish law, which enforced Jewish separation as a means of "self-preservation." This ethic bred a "Jewish racial instinct" that justified a mass Jewish return to the Middle East. Auerbach's essay was soon followed by physician Felix Theilhaber's startling *Der Untergang der deutschen Juden—The Demise of German Jewry*—which spoke of race suicide caused by conversion and a general indifference to Jewish culture among Jews. Prominent Jews even hinted that racism might have helped preserve Jewish identity. "It may be thanks to anti-Semitism that we are able to preserve our existence as a race," said Albert Einstein in 1920.

Though eugenics and racial chauvinism are offensive by today's sensibilities, this strong tide of hereditary thinking crossed oceans and ideological barriers. The scientists, politicians, and celebrities who espoused genetics-based solutions to social issues were, by and large, respected and respectable. In a famous *Harper's Magazine* article published in 1899, Mark Twain noted with some amazement that world Jewry, but 0.25 percent of the human race, was "a nebulous dim puff of stardust lost in the blaze of the Milky Way. Properly, the Jew ought hardly to be heard of; but he is heard of, has always been heard of. He is as prominent on the planet as any other people, and his importance is extravagantly out of proportion to the smallness of his bulk . . . What is the secret of his immortality?" he asked.

According to the popular wisdom, good breeding was the answer. Many prominent thinkers, conservatives and liberals alike, including H. G. Wells, Emma Goldman, Alexander Graham Bell, George Bernard Shaw, Winston Churchill, and even Margaret Sanger, the founder of Planned Parenthood, enthusiastically embraced what became known as "positive eugenics." Famous eugenicists included the Reverend Harry Emerson Fosdick of Manhattan's liberal Riverside Church, Baptist preacher Russell H. Conwell, and Rabbi Stephen Wise of the Free Synagogue of New York, the most influential Reform rabbi in America in the early twentieth century. Another New York rabbi, Max Reichler, in a famous essay on Jewish eugenics published in 1916, claimed that the systematic adherence to the basic principles of Jewish religious law was the "saving quality, which rendered the Jewish race immune from disease and destruction." In a Mother's Day sermon delivered during the 1920s, the Kansas City rabbi Harry Mayer went so far as to declare, "May we do nothing to permit our blood to be adulterated by infusion of blood of inferior grade."

In western Europe and the United States, such racializing eventually became indistinguishable from xenophobia, sterilization laws targeting "inferior breeds," and restrictive immigration laws—all expressions of "negative eugenics"—ridding the human race of its moral and physical detritus. The Johnson-Reed Act of 1924 sharply restricted immigration of the "huddled masses" yearning to escape desperate poverty in southern and eastern Europe, which effectively ended fifty years of large-scale Jewish immigration to the United States. In Nazi Germany, eugenics was used to justify transforming science into a handmaiden to the state, with race purity as its driving ethic. Charles Darwin's theory of natural selection had by that point been twisted far beyond recognition. Race no longer designated a population group that shared physical traits. Instead, it had become a way to stigmatize "anyone but us." The bright stream of evolution had turned into a raging river of racism.

ZIONISM AND THE RIGHT OF RETURN

Zionism was forged during this era of intense nationalism and racial chauvinism. Blood ancestry—a vague term to biologists but a very real one to ethnic groups consumed with their identity—had proved problematic for Jews. But it also defined Israel. To most Jews, Israel was a blood link that had persisted through crises of history, an idea as much as a country, a historic mission and a national identity. That's why Zionism, which started as a movement of mostly secular Jews, also had resonance with religious Jews. It is seen by believers as the embodiment of God's promise to restore the Israelites to their homeland, a connection first severed when Jews were exiled to Babylon more than 2,500 years ago. "By the water of Babylon," reads Psalms 137:1, "there we sat down and wept when we remembered Zion."

Even before the emergence of Zionism in the late nineteenth century, Ottoman Palestine had a small but substantial Jewish population consisting mostly of Sephardic refugees from Spain and a trickle of immigrants from other Middle Eastern countries. In 1882, Russian émigrés fleeing pogroms founded the first Zionist settlement, Rishon LeZion, which literally means "the first for Zion." At first, most western European Jews remained either cool or openly hostile toward Zionism; assimilation seemed a far more promising solution to the periodic anti-Jewish flare-ups that had defined Jewish history. That attitude changed with the Dreyfus Affair.

In 1894, Captain Alfred Dreyfus, the son of a prominent Parisian Jewish family, was arrested, charged with treason by the French rightist government for supposedly smuggling documents to Germany, convicted, demoted, and exiled to Devil's Island off the coast of South America. He was later freed subsequent to appeals (and eventually, in 1906, found innocent). The ordeal traumatized European Jewry. A Jewish journalist from Austria, Theodor Herzl, who covered the trumped-up trial and the hysteria that accompanied it, was shocked that Jew hatred was so ingrained in the "civilized" French. Secular and assimilated, Herzl came to believe that the only safe place for the Jews was a land of their own.

Shortly after the trial, he wrote *Der Judenstaat—The Jewish State*—which made the political case for Zionism, which at the time was a mostly Russian secular movement that had sprung up in Odessa. Why have Jews historically found themselves enveloped by separateness, objects of distrust and sometimes hatred and persecution, even in communities in which they have flourished, he asked. "I consider the 'Jewish question' neither a social nor a religious one, even though it sometimes takes these and other forms," he wrote. "It is a national question, and to solve it we must first of all establish it as an international political problem to be discussed and settled by the civilized nations of the world in council. We are a people—one people." What soon would become a brand of death for Jews—claims of Jewish specialness—was, in the years before the Holocaust, cause for revivifying an eroding identity.

Herzl convened the First Zionist Congress in Basel, Switzerland, in 1897 as a symbolic parliament for those in sympathy with the still-remote idea of the return of Palestine to the Jews. The meeting drew a motley collection of 197 Jews of all religious and ideological stripes: Orthodox, nationalist, liberal, atheist, anarchist, socialist, and capitalist. For the core leadership made up of mostly liberal, secular Jews, Zionism marked a departure from the dalliance with assimilation and an affirmation of Jewish peoplehood. "Zionism is a return to the Jewish fold even before it is a return to the Jewish land," Herzl declared. He grounded the movement in a cause even secular Jews could embrace: the biblical manifest destiny of Eretz Yisrael, Jewish land, a gift of exceptionality supposedly given permanent stamp by divine commandment.

DIVIDED PALESTINE

The currents of nationalism that ran through Europe and encouraged the Zionist dream also spurred hopes among the Arab population in Palestine, which had long resented the political domination of the Ottoman Turks. Arabs did not arrive in the Holy Land in large numbers until after Muhammad's death, in 632, but they became a dominant presence and were the overwhelming majority at the turn of the twentieth century. As a reward for acts of sabotage against Turkey, the Arab Hashemites believed they would be granted rule over an Arab kingdom. But as it turned out, commitments made in the heat of battle would not determine the borders nor the identity of the modern Middle East as much as political agreements hammered out between Western powers would. After World War I, the victorious British and French descended like packs of hungry wolves on the rotting carcass of the Ottoman Empire. The victors rebuffed any plan to establish an Arab kingdom or federation of Arab emirates.

The British were guided in their maneuverings by the Balfour Declaration. Arabs had been stunned when on November 2, 1917, the British foreign secretary Arthur Balfour sent a letter to the Zionist leader Walter Rothschild, a former Liberal Party representative in Parliament and member of the famous Rothschild banking clan, setting out Britain's revised policy on the future of Palestine. "His Majesty's Government view with favour the establishment in Palestine of a national home for the Jewish people," Balfour declared, although he also wrote that "nothing shall be done which

Figure 10.1. The dismemberment of Ottoman-ruled Palestine.

may prejudice the civil and religious rights of existing non-Jewish communities in Palestine." Arthur Koestler later famously wrote that the Balfour Declaration was a case in which "one nation solemnly promised to a second nation the country of a third."

The Zionist dream was generally viewed sympathetically in the West. "The four Great Powers are committed to Zionism," Balfour wrote two years after his famous letter. "And Zionism, be it right or wrong, good or bad, is rooted in age-long traditions, in present needs, in future hopes, of far profounder import than the desires and prejudices of the 700,000 Arabs who now inhabit that ancient land." Zionists also drew strong support from Christian millennialists, particularly in the United States, where restoring Jewish domination in Palestine was long seen as a prelude for the return of Jesus. In 1863, Abraham Lincoln had underscored the popular sentiment when he declared, "restoring the Jews to their homeland is a noble dream shared by many Americans." George Bush, a New York University Bible scholar and the progenitor of two presidents, had actively campaigned for a Jewish state in Palestine.

Restorationism, as it was called, reached its zenith in the Blackstone Memorial, a petition submitted by midwestern businessman William Blackstone and cosigned by fellow magnates John D. Rockefeller, J. Pierpont Morgan, Charles Scribner, and William McKinley. It urged President Benjamin Harrison to convene an international conference to discuss ways to revive Jewish control over the Holy Land. "It seems to me that it is entirely proper to start a Zionist State around Jerusalem," wrote Teddy Roosevelt, "and [that] the Jews be given control of Palestine." President Woodrow Wilson saw the possible return of the Jews to Palestine as the fulfillment of biblical prophecies, and he fervently backed the British Zionist project.

Understandably, British contemplation of a possible Jewish state compromised Arab expectations of establishing a homeland based in Jerusalem. Britain assumed authority over Palestine in 1920 (ratified by the League of Nations in July 1922). With Palestine's future in limbo, the Hashemite Faisal was left to scramble for what remained. After negotiations with the French, he was declared king of Syria in 1920, but after being granted control by the League of Nations, the French snatched the country back in a bloody battle. Faisal departed to exile in London. In order to placate Arab dis-

our DNA can reshape how we think of ourselves as members of races, ethnic groups, and religions.

King was a rising star while still a graduate student at the University of California at Berkeley in the 1970s, when she helped establish that humans and chimpanzees were, genetically speaking, kissing cousins. Her early work helped lay the groundwork for the discovery of the "First Mother"—Mitochondrial Eve. But she is best known for helping first identify the gene that carries the mutation that causes breast cancer, one that disproportionately targets Jewish women. It's one of dozens of genetic disorders that are popularly known as "Jewish diseases."

King's early successes offered her the luxury of setting up her own laboratory at Berkeley, stocked with crackerjack graduate students, to fulfill her dream of finding a cure for breast cancer, one of nature's most wanton killers. The disease strikes one of every eight women. A diagnosis presages a frightening and often savage struggle for survival. Every twelve minutes or so, 45,000 times a year, an American woman dies of the disease. At any one time, a million women in the United States are unaware they are affected. There are 215,000 new cases diagnosed annually; tens of thousands more women are victimized by ovarian cancer.

When she embarked on what many told her was a quixotic effort, no one was sure what mix of genetic and environmental factors might cause female cancers. In those days, it was considered a long shot that genetic patterns would even be identified in families with a history of one of these killer diseases. It would be a few years before scientists would finally be able to demonstrate that defective genes gone haywire provoke cells to recklessly divide until mutant tumors ravage their hosts.

King focused on trying to come up with a sentinel system to warn patients and their doctors that a disease may be lurking in otherwise healthy young women. She focused her research on two dozen families in which breast or ovarian cancer victimized related sisters, daughters, mothers, grandmothers, or aunts. She remembers the lonely days in the 1980s when she wondered if she had turned down a blind alley in what would become a fifteen-year quest. But the drudgery in the lab gradually began to pay off.

"It was like solving a puzzle," she said. "You could see it in the data. Family history was one of the best predictors of breast cancer. One day a light went on. We realized that the gene, if it existed at all, was likely to be found by focusing most especially on families where cancers struck young women, many by the time they were forty-five years old." It was not the breakthrough she had been hoping for, but it was a start.

Though King and her team were encouraged by their early findings, things turned ugly outside the lab. Skeptical and jealous colleagues began whispering that she might be fabricating results to promote her career. "It was very discouraging for a while," she recalled with pain in her voice. "I was under constant attack. I was concerned how the pieces of our research would come together, and when."

Finally, after years of painstaking effort, in 1990, the breakthrough came. By sorting through the DNA of each patient in the extended families she was researching, she identified a cancer perpetrator in the genome. It was on the seventeenth of the twenty-two chromosomes. King christened it Breast Cancer One, written as BRCA1, although the mutation also causes ovarian and even prostate cancers and can be passed along by and to both men and women. She had solved one of the great mysteries of the dawning age of the genome. With this discovery, King moved from renown to outright celebrityhood. She was featured in the *New York Times* and *People* and named by *Glamour* magazine as one of the Women of the Year.

Finding the general location of the gene made fame and headlines, but the money was in locating its exact spot on the genome. It fired a cloning research competition between the King lab and other geneticists—a mad dash that the media covered like a horse race. In 1994, a team at a Salt Lake City research company, Myriad Genetics, isolated BRCA1, exactly where King predicted it would lie. Each mutation in BRCA1 is an infinitesimally small "mistake," but enough to hamper the gene's natural function—its ability to suppress cancer tumors. Eventually thousands of different mutations were found on BRCA1 and another breast cancer–carrying gene, BRCA2. Together, the defective genes are responsible for at least 5 to 10 percent of all breast cancers and 15 percent or more of ovarian cancers in American women.

But there was another twist to the story. As the case histories began to pile up, the cancers turned up most commonly in a strikingly tiny fraction of the population: Jews. On average, women have less than a 0.1 percent chance of having one of these genes. Subsequent studies found that about one in eight hundred women and men in the general population carries one of the three mutations. In contrast, approximately one in forty Jews (2.5 percent) is a carrier, extraordinary for a cancer-producing gene. Women with mutations in BRCA1 or BRCA2 are five to fifteen times more likely to develop breast or ovarian cancer than the rest of the female population. Mutations in BRCA1 and BRCA2 can raise lifetime breast cancer risk to as high as 85 percent and the chance of developing ovarian cancer to as high as 50 percent.* For many women with these mutations, the deadly question is not whether but when the mutinous genes will finish their work.

"JEWISH DISEASES"

Medicine and disease have been peculiar tropes of Jewish identity since biblical times. Jews have often been admired for their medical prowess and yet stigmatized as inferior or defective, morally and physically. The popular image of the Jewish doctor has roots in Isaiah, who cures King Hezekiah. Jesus healed through faith. Jews later

---

* BRCA1 and BRCA2 mutations have been identified in all races, including blacks. Among African Americans, some BRCA1 and BRCA2 mutations are of European origin and others can be traced to specific regions of Africa that were sources of the slave trade. As in Jews, cancer tumors caused by the mutation are likely to be aggressive and deadly.

developed a strong tradition of folk medicine, drawing on Greek and Roman texts. During the diaspora, Jews focused on healing as a religious calling, which gave medicine a special status within Jewish culture. The legendary image of the Jewish doctor arose in part out of the differing Jewish and Christian pre-Enlightenment views of medicine. While the medieval church was focused on otherworldly matters, attributing disease to supernatural causes and considering the body a vessel of evil, Jews took a more commonsensical view—they became doctors.

Moses Maimonides, revered as a physician and court attendant to the ruler of Egypt in the thirteenth century, is memorialized in Jewish history as a medieval Albert Schweitzer, but he was hardly atypical. From the tenth century onward, Jewish physicians were recognized for translating ancient medical texts, writing medical treatises, and practicing medicine in the noble courts of Europe. According to a sixteenth-century anecdote, Francis I of France, suffering from a lingering illness, had the Holy Roman Emperor, Charles V, send him his Jewish physician. Apparently attempting to ingratiate himself with King Francis, the doctor told him that he was no longer a Jew, having been converted to the only true faith—whereupon the indignant king immediately dismissed him and asked for a real Jewish physician! As the Enlightenment flowered, Jews in greater Germany and in other liberal regions in Europe began streaming into the universities to study modern medicine.

Brainy, yes, but Jews could never shake the Christian stereotype that they were otherwise congenitally enfeebled. It's probably no coincidence that Moses Mendelssohn, hunchbacked and speaking German with a Yiddish accent, was chosen by the German establishment as the pet Enlightenment Jew. He visibly and conveniently embodied many of the racist stereotypes of what a Jew should look like. Even as Jews successfully integrated themselves into secular German society, the notion of Jewishness was associated with illness and disease, from diabetes to mental disorders such as depression and schizophrenia. Anti-Jewish nationalists viewed these and other disorders that seemed to show up commonly in Jews as proof of the fundamental degeneracy of the "Jewish race." It's not hard to trace the line from these stereotypes to the Final Solution.

In fact, this racist view was grounded in a sad reality. Because of their cohesive history, Jews do suffer disproportionately from dozens of disorders. The majority of these diseases are recessive and linked to mutations on the autosomes, the twenty-two nonsex chromosomes (the X and Y chromosomes are called sex chromosomes, as they determine our sex). It was speculated at first that they might have arisen solely because of inbreeding in traditionally close-knit Jewish communities. But if that had been the only explanation, the victims of these disorders would have had fewer children or died out over time, in effect naturally selecting these disease-causing traits out of the gene pool. They are now attributed to a combination of inbreeding; chance mutations as the result of genetic drift; the founder effect, in which the population of European Jews passed through various bottlenecks; and possibly positive selections in which a sometimes deleterious trait is preserved because it also confers some evolutionary advantage.

Because of their history of cultural insularity, Jews are more likely to be targeted by diseases caused by recessive genes, but overall, they are no more likely than any other ethnic group to be born with a genetic disorder. Everyone, regardless of ancestry, is walking around with as many as fifty significant glitches in his or her DNA. "It just so happens that we happen to know more about what the risks are in the Jewish population," said Harry Ostrer, a pediatric geneticist and the director of the Human Genetics Program at New York University Medical Center. Ostrer has estimated that one in three hundred Jewish children is born with one or more of these diseases, and Jews are a much higher percentage of carriers.

Although the term "Jewish disease" is widely used, even among scientists, many disorders that are unusually common among Jews—the cancers, Parkinson's disease, cystic fibrosis, hearing loss, and others—show up in non-Jewish populations, though usually in smaller percentages. [See appendix 5 for lists of "Jewish diseases."] "We now know that most, if not all, human disorders have a genetic background, and we're acquiring the tools to study, treat, and eventually prevent or cure them," noted Gideon Bach, head of the Department of Human Genetics at the Hadassah-Hebrew University Medical Center at Ein Karem in Israel. "It's far easier to trace genetic anomalies in groups with homogeneous pedigrees [like Jews]."

Most Jewish diseases are thought to result from mutations that have appeared since the founding of Ashkenazi Jewry, over the past thousand years, and have been preserved because of that group's history of cohesiveness, at least until recent decades. They fall into two major categories: breast, ovarian, and prostate cancer, and other diseases, such as Bloom syndrome and Fanconi anemia, that result from breaks in the system that repairs damaged DNA; and an astonishing number of rare brain and nervous system disorders, such as Gaucher, Niemann-Pick, and Tay-Sachs, that also may hold clues to the mystery of intelligence.

There are fewer genetic disorders specific to Sephardic Jews, Oriental Jews, and other small Jewish populations, probably because they intermingled more with gentiles than did European Jews. Consequently, they often manifest the same genetic disorders that occur in the non-Jewish population of their native countries. Unique diseases also show up in two small populations with Jewish roots that have caught the attention of geneticists: the Samaritans and the Karaites.

Samaritans suffer from many genetic disorders, some linked to their ancient roots and so shared by their Middle Eastern neighbors, and others, like color blindness, that are of more recent origin and relatively distinct to them. Jews, Samaritans, and indigenous Arabs in Israel are regularly tested for the genes linked to congenital deafness and a muscular disorder that can make it troublesome to walk. "When young people want to get married, nobody wants to think about genetics," said Benyamim Tsedaka, a Samaritan who often speaks for the community. "But we have learned to be more careful."

The Karaites are a splinter group of Jews who broke away from the mainstream Jewish community of Babylon in the fifth century in a rebellion over its bureaucratic rule and high taxes. Originally known as Ananites after their leader, Anan ben David, a member of an aristocratic Rabbanite family in Baghdad, they rejected rabbinic Judaism and its reliance on the oral law that had evolved over the centuries, and insisted that only the five books and the book of Psalms be read in the synagogue. Following ancient tribal tradition, men and not women passed on their religious identity to their children. Anan's followers later took on the name Karaites and saw themselves as the true People of the Book. Like the Samaritans, the Karaites married only within their community, but unlike many other Jewish communities, they blended in culturally with their neighbors. Because of this, and although they were no less biologically distinctive than other Jewish communities, nineteenth-century Russian authorities and later Nazi-controlled Lithuania designated Karaites as a Jewish sect by religion and not a "race." That spared them anti-Jewish legislation and in some cases their lives.

Karaites from the Soviet Union, Syria, and Egypt were eventually accepted as citizens of Israel under the Law of Return, although they were classified as "non-Jewish" after outcries from Orthodox rabbinic officials. European Karaites appear to trace their genealogy to the Karaite community of Istanbul. Mutation studies have found that Karaites share a number of genetic disorders with Middle Eastern Jews and Arabs. However, suggesting that their claims (and the beliefs of Russian and Nazi authorities) of ethnic purity are overblown, DNA studies also indicate considerable intermixing over the years with the Tatars and other groups among whom they lived for centuries. Today, only about ten thousand descendants of this ancient sect remain.

TAY-SACHS

The discovery of dozens of disorders unusually common among and sometimes almost exclusive to Jews has presented a vexing challenge to the Jewish community. One reporter recounted a gathering of some of Boston's most prominent Jews for a quiet, off-the-record discussion about "Jewish genes." Over sandwiches and cookies, they learned of the latest research about the breast cancer gene that Mary-Claire King had originally isolated. Judy Garber, a researcher at the Dana-Farber Cancer Institute, posed this question: would their organizations contribute to future studies of the genetic distinctiveness of Jews that might help save the lives of many Jews and further genetic research? To Garber's surprise, the answer generated universal hand-wringing and many polite maybe's and no's.

"The feeling was, 'Why us?'" said Nancy Kaufman, the director of the Jewish Community Relations Council. Those present expressed concern about targeting Jewish women or of affirming aging stereotypes that Jews are more prone to disease. They worried that this renewed interest in ethnic and racial differences reflected a problematic resurrection of the notion that Jews were a distinct and inferior racial group.

The possible misuse of disease research has not discouraged scientists; the potential health and medical benefits are just too immense to ignore. The study of "Jewish genetics" has already saved thousands of lives around the world. Consider the scourge of Tay-Sachs, an inherited disorder of the central nervous system. It has been nearly eradicated in the United States. This success story—one of many more to come, geneticists hope—began tragically, with the death of a beautiful baby boy in 1969.

"For the first six months of his life, he seemed perfect," said Rabbi Josef Ekstein, a leader of the Brooklyn Lubavitcher, the community of Hasidic Jews known for their traditional Orthodox dress and commitment to religious study. "Then I noticed he was losing some of his capabilities. He got all floppy and couldn't support himself. I started to run around asking doctors what was wrong and they said: 'He's a lazy child. Don't worry about it.' As the days passed, the boy's health worsened. He started having seizures, he couldn't move, he couldn't swallow and eventually he went blind."

When their son was two years old, the Eksteins took him to the world-famous center for the study and treatment of genetic nervous disorders at Mount Sinai in New York City. "He was diagnosed with Tay-Sachs," Ekstein recalled with anguish. "This was the first time I'd heard of it. My son suffered terribly and died at the age of 4." For the Eksteins, the nightmare was just beginning.

Tay-Sachs was first identified in 1887. Bernard Sachs, a New York neurologist, found himself helpless in treating a Jewish infant with severe mental problems. The boy had been born healthy and for months developed normally. But by the time he was a year old, he was blind, deaf, and unable to swallow. His muscles had turned rigid, and his head began to swell. Sachs put him through a battery of tests at Mount Sinai Hospital on New York's Upper East Side, but nothing could be done, and he soon died. Other parents brought infants with similar problems to him, but he was helpless to do anything. Each of them suffered a horrible death. And each had a telltale splotch: a very distinctive cherry red macula in the eye.

Convinced the disorder was inherited, Sachs named it "amaurotic familial idiocy." Only later did he find out that a British ophthalmologist, Warren Tay, had published an account of a similar case in 1881. Later it was recognized that Tay and Sachs were seeing different facets of the same condition, which is caused by the buildup of harmful quantities of a fatty substance in the nerve cells in the brain.

It was not until the last quarter century that anyone realized it might be possible to contain Tay-Sachs. Because marriages within the ultra-Orthodox Jewish community are generally prearranged and are sometimes between cousins, genetic diseases are common. Twenty-five years ago, every Orthodox Jew knew of someone who had been victimized by Tay-Sachs. In Crown Heights, where the Eksteins live, an estimated one person in sixteen carries a mutation for the disease, about one hundred times more common than in the general population.

By the 1970s, researchers had developed a screen, but for ultra-Orthodox Jews, abortion and prenatal screening are strictly forbidden. "A disease that runs in the family was a very taboo subject in our community," Ekstein said. "Families who had children with diseases felt stigmatized and didn't talk about it for fear that their healthy children would not be able to marry. I know of two brothers who had children with Tay-Sachs, but neither told the other. I heard of a doctor and his wife who had a Tay-Sachs child and would hardly ever go out. Many ill children were sent away to care homes. I did this myself. I was one of those people who had tried to deny the problem."

The family believed they could do nothing but hope and pray, but that didn't work very well. Over the next eighteen years, they watched three more of their children waste into retardation and death until they could follow the community's narrow dictates no longer. "I didn't want another person or another family to go through that," he said. "No matter what, this had to stop."

There was great resistance when Rabbi Ekstein first proposed conventional testing. But a year later, in 1983, he devised an ingenious plan that did not directly violate his community's strict religious beliefs and avoided the stigma and discrimination that could occur if someone was found to have the Tay-Sachs mutation yet have no sign of the disease, and the diagnosis became public. Under the screening system known as Chevra Dor Yeshorim, children are tested well before they are of marriage age—at schools, synagogues, community centers, wherever. No names are collected; everyone is issued a coded number. The Lubavitcher soon embraced the plan with the same fervor it reserved for High Holy Day services. During courtship, before a couple is engaged, Dor Yeshorim now informs them whether they are genetically incompatible—that is, if both carry a mutation for Tay-Sachs. They are not told if only one person is a carrier, since the couple would have no chance of having a Tay-Sachs baby, and the mutation otherwise has no effect on health.*

The screening program has been extraordinarily successful. So far, more than 250 prospective husbands and wives have broken off their courtship after they were both found to carry the mutation. In the 1970s, Tay-Sachs cases ran eight to one Jewish to non-Jewish. The sixteen-bed Tay-Sachs ward in New York's Kingsbrook Jewish Medical Center in Brooklyn was filled to capacity. Since 1996, the ward has had no Tay-Sachs patients. The number of babies with the disease has dropped from fifty a year to five—and to almost none from the Orthodox community, where testing is widely embraced. Genetic screening for Tay-Sachs, the prototypical "Jewish disease," has been so successful that the disease now turns up more frequently in non-Jews.

The program's success has led to the development of similar screens for more than a dozen other Mendelian recessive diseases— illnesses caused by the matching of single recessive genes from each parent—common among Ashkenazi Jews, including Canavan disease, a degenerative disease that causes weakness, seizures, and eventually death (1993); Gaucher disease, which can result in anemia, fatigue, bone degeneration, and neurological problems (1994); mucolipidosis IV, which is marked by progressive retardation (2000); and familial dysautonomia, a disorder that affects the senses and involuntary actions such as digestion, breathing, and the regulation of blood pressure and body temperature (2001). Although these diseases have not been eliminated, the combination of screening and new treatments has allowed sufferers of some disorders, like Gaucher, to live a relatively normal life. Ostrer has estimated that 1 in 4 Ashkenazim carries a mutation for one of these conditions, though only a tiny fraction of the carriers will develop

---

* If only one parent is a Tay-Sachs carrier, there is no chance of having an affected child and a 50 percent chance the child will be a carrier. If both parents are carriers, their child has a 25 percent chance of being disease-free and not carrying the gene, a 50 percent chance of being an asymptomatic carrier, and a 25 percent chance of certain death by the age of five.

the disease, because it is recessive. Without screening, there's a 1 in 368 chance that a carrier marrying an Ashkenazi Jew will have a child with one of these diseases.

The outreach program pioneered by the ultra-Orthodox community in Brooklyn has been extended to Jewish communities across the United States, Europe, Australia, and Israel and has become a model for genetic testing worldwide in communities where marriage between close kin is common, such as in Arab cultures. Millions of people have now been tested using this system, saving countless lives. DNA disease testing services have sprung up, such as HealthCheckUSA, which for $200 and up will tell you whether you carry the mutation for eight genetic diseases. Ironically, one major pocket of indifference has been the mainstream American Jewish community, which has been reluctant to back funding initiatives, perhaps because of the historical stigma of the "defective Jew." "The organized Jewish community has been slow to step up to the plate about it," Ostrer said. "We've tried to generate interest, and it has fallen flat." But intermarriage, while draining Judaism of its ancestral distinctiveness, may be slowly addressing the scourge of Jewish diseases. "For people who report Jewish ancestry, we're seeing that the frequency of specific mutations appears to be dropping over time," he said. "People have mixed ancestry. They're three-quarters Jewish, not fully," Ostrer added.

KING'S JEWISH ROOTS?

Mary-Claire King's groundbreaking discovery has helped deepen a very personal and highly unexpected connection she has developed with Jews and Judaism. She grew up in Wilmette, a prosperous suburb north of Chicago. Harvey King, her father, was a personnel manager for Standard Oil. Her mother, Clarice, worked for the War Labor Board until she became pregnant with Mary-Claire. When she was a young girl, her mother would regale her with tales of her Pilgrim heritage. The Gates family, her mother told her, could trace their lineage back thirteen generations to the original Plymouth community, in Massachusetts, and to England.

"Our ancestors were Separatists who believed the Church of England had become corrupt," she said. By the early 1600s, the Separatists had begun forming secret congregations, convinced the church couldn't be purified of what they considered false ceremonies, non-Scriptural teachings, and superstitious rituals. During the reign of Queen Elizabeth I, Puritan-leaning religious opinions were generally tolerated. When King James succeeded her in 1603, he began to prosecute, fine, and jail those who openly defied the church. By 1620, religious persecution and the dismal economic situation led the Separatists to abandon Europe for a fresh start in America.

With the help of merchant investors, the Pilgrims got a charter to start a new colony in Massachusetts. One Separatist group left Holland on the *Speedwell*, a small ship, and sailed to England to join with the *Mayflower*, a large cargo ship that had been most recently used to ferry wines in from France. Together the ships would sail to America. The Ring family was on the *Mayflower*. "The family, at the time of the beginning of the voyage in 1620, consisted of a father, William Ring, a mother, and three little girls," King said. But on the voyage, William died, throwing the family's plans into disarray. The ship's captain had the Ring family, now fatherless, put on the *Speedwell*, which by that time had developed a leak and had to return to England.

Widow Ring, fearing for the safety of her family in England, continued to hope for a second opportunity to come to America. But the chances were dim. "No one was allowed to come without a head of the family, without a man to support them," King said. "But the widow absolutely insisted to be brought even though she was not married." King breaks into a smile, thoroughly identifying with her ancestor, a feisty widow. She herself is a single mother, having raised her daughter, Emily, after a divorce more than twenty years ago. "There were letters back and forth between her and the governors of the community. She said, 'I'm thoroughly capable of taking care of myself.'" And off Widow Ring went to Massachusetts.

In King's mind, growing up in suburban Chicago, she was pure Yankee, with roots sunk deep into the soil of the Massachusetts Bay Colony. That confident world turned upside down during Christmas in 1966. She had just returned home from Berkeley, where she was studying statistics as an undergraduate. "I had taken all my worldly goods to college in the fall. So, I didn't have any books at home. And I needed a dictionary. I went to my mom's bookshelf. I pulled down a dictionary from, it must have been about 1934, and flipped it open to look up whatever word it was. And there, in what was clearly my mother's handwriting, up at the top, was written 'Clarice Cohen.' The very first thing that went through my mind was, 'Oh, it isn't my mother's dictionary, and it belonged to some other person named Clarice Cohen.' Then I thought, 'Clarice is a pretty unusual name.' The second thing that flashed through my mind was that she must have been married before and not told me. So of course, I was curious. I carried the dictionary into the kitchen, where my mother was washing dishes. She was totally unprepared for me to say anything except 'Hi.'"

"You never told me you were married before," she remembered saying to her mother, who rattled the dishes in the sink, she was so taken aback by the question. "'I was never married before,' she said. 'What on earth makes you think that?' I held up this dictionary and said, 'Oh, because your name used to be Cohen.'

"She hadn't turned around yet; she was still washing dishes, and the first thing she said, as she turned around, was, 'My name is not Cohen.' I could hear in her voice that she was scared.

"I was holding her Webster's dictionary. I opened it up and said, 'It says Clarice Cohen right here.' And she dropped a glass she was holding. 'You mustn't ever tell. You mustn't ever tell anyone,' she begged. 'It's really not safe. Don't ever let anybody tell you it's safe. You must never tell.'"

King was baffled by the palpable fear that rose from her mother like heat off a skillet. "I said, 'What are you talking about?' After that, she started to cry, and it all just came out."

King recalled being dumbstruck. While Separatist and Methodist by family history and religion, King was part Jewish by genetic ancestry. Her mother's father, Louis Gates, was originally named Cohen. "I think back now and I can picture him. Although I had never thought about it before, he looked just like a sweet pixy Jewish grandfather. If you said, 'Draw me a Jewish grandfather,' it would be my grandfather. But I knew him from the time I was a tiny child as Louis Gates." Her grandfather could by oral tradition trace his ancestry to the time of Moses and Aaron. Here she had grown up believing she was a Colonial Dame of America only to find out she was part Jewish American Princess.

Louis Gates was born Louis Cohen in New Orleans, where his parents had settled after arriving from eastern Europe in the 1880s. Like many Jewish families living in the South at that time, the children strived to blend in with their Christian neighbors in their adopted homeland. As a young man, Louis took a job as a department store manager in Saint Louis, where he met and married Clarice's mother, who was also named Clarice. She was indeed a Pilgrim-descendant and a member of the Christian Church. As Louis was not religious, the children were raised as Christians, although the family name remained the very Jewish-sounding Cohen.

"They were not well off by any means, but they were not in terrible straits," King recalled her mother telling her. Opportunity beckoned in Tulsa after World War I, and they resettled. Tulsa's tiny Jewish community consisted of fewer than a dozen families. "My mother's family thought of themselves as Christian, but nevertheless they were perceived as Jewish by people in town." Then two traumatic events sealed the family's fate as Jews, whether they sought that identity or not.

The first occurred in May 1921. Nineteen-year-old Dick Rowland got into an elevator in the Drexel Building in downtown Tulsa. He accidentally stepped on the foot of the elevator operator, Sarah Page, a white woman. That was not a good turn of events for a young black man. Not knowing it was accidental, she attempted to hit him with her purse. Rowland grabbed her, she screamed, and he ran out of the elevator and the building. Page told the police that Rowland had tried to criminally assault her, although she later changed her story and said he only grabbed her. But when the *Tulsa Tribune* published an account that afternoon, it was reported as an assault. By nightfall, it was rumored that Dick Rowland, now being held in the county jail, could measure the rest of his life in hours.

Throughout the next day, crowds of whites gathered outside the jail. By nightfall, with more than four hundred people gathered, seventy-five local blacks, some armed, came to the jail to prevent what they feared was going to be a lynching. Apparently, a white deputy attempted to disarm one of the blacks, a shot was fired, and all hell broke loose. Mobs of whites began to drive around the streets, shooting any black person they saw. The governor called out the National Guard. By morning, a mob of fifteen thousand whites marched on the Greenwood District, also known as Little Africa. They used machine guns and dropped nitroglycerin in an all-out attack, killing, looting, and burning everything in sight. Martial law was decreed as the National Guard took control of policing the city. Eventually more than 6,000 rioters were apprehended. The destruction was staggering: 300 dead; 1,200 homes destroyed; 10,000 blacks left homeless. By the end of that year, the Ku Klux Klan had attracted more new members in Oklahoma than were thought to have existed in all of the rest of the country only the year before.

The riots and their aftermath were a defining event for Tulsa and the Cohen family. As the riot unfolded, with the city still on fire, some white families rushed into Little Africa to see what they could do to help. The Cohens joined them. "My mother must have been six when that happened. They got in their car, drove across town, brought people out of Greenwood, and hid them for months in their houses. There were many people whom her mother and father knew and cared about who were Negroes who were burned out of their homes."

Overnight, the Cohens became strangers in a suddenly strange land. Although other Christian families along with the Cohens had gone to the aid of local black residents, "the response to them had everything to do with their name being Cohen," King recalled her mother saying. "The response was that the Klan came after them. They lived in terror of their house being burned down or worse. I think it colored her years as a child because the people she was going to school with were children of these Klan people."

The second trauma occurred a decade later, as Clarice Cohen applied for college. King remembered standing in the kitchen that morning as her mother told her, for the first time, her heartbreaking story of being denied admission because of her last name. Clarice was the top student in her class, yet she was rejected from all of the Eastern Ivy and sister schools to which she applied. "'Dear Miss Cohen,'" read one stinging letter, according to King. "'Your application is absolutely outstanding. There is no question that it is at the same level as the other members of our entering class, but we have already filled our quota on Jewish young ladies.' And that was that," King said, as she recalled the anguish in her mother's voice. "The fact that she wasn't Jewish was irrelevant."

In the 1930s, the "one-drop rule" applied to both blacks and Jews. Clarice Cohen went instead to the tolerant and intellectually exhilarating University of Chicago. But she was forced to return to Tulsa after two years. "The Depression was getting worse and worse and worse. Jews were fired first, which meant her father lost his job." Clarice then watched helplessly as her mother died in her early forties of cervical cancer that went untreated because she could not afford medical care. "Clarice landed a job and paid off the family debt. Then she legally changed her family's name, including her father's, from Cohen to her mother's maiden name of Gates. She took her new name, she went back to Chicago, and she never looked back."

King had identified with her mother's anger at the injustice that had befallen her, but the fear she heard in her mother's voice long perplexed her. "I didn't fully appreciate it when she first told me," King said. When she was in high school, King had often dated Jews, which drove her mother crazy. "She would say, 'I don't want you to go out with Jewish guys.' Of course, I read it as anti-Semitism on her part. And until recently, each time I've come up for a job that involved FBI scrutiny, my mother would say, 'What if they find out we're Jewish?' She thinks that anti-Semitism in America is just one layer below the surface. She thinks it's everywhere. She thinks the Klan is still here. They just don't wear sheets anymore."

King settled back in her chair and took a breath. She had been talking almost nonstop for the better part of an hour. My next question, one I would have thought she would have anticipated, caught her off guard, which by then I had thought was impossible to do. Had the discovery of her hidden ancestry, her Jewish ancestry, had any impact on her own identity?

In fact, there had been a subtle but sure shift in her career trajectory after the surprise revelation. Shortly after she had returned to Berkeley, she refocused her studies from statistics to genetics. Her rediscovered Jewish roots have never been far from her thoughts. "I've gotten particularly involved in doing genetics in the Middle East with Israeli and Palestinian colleagues," she told me. In 1996, she launched a study, ongoing, to find the origins of congenital deafness among Jewish and Arab children. That same year, she teamed with Joan Marks, a social worker and the founder of genetic counseling, to begin the New York Breast Cancer Project. They recruited more than one thousand Jewish families with breast cancer from twelve New York area centers and carried out the definitive study of the risk of breast and ovarian cancer in women carrying mutations in BRCA1 or BRCA2. The results yielded a higher than 80 percent risk of breast cancer by age seventy, compared to 12 percent for the general population. The study revealed a risk of ovarian cancer as high as 50 percent.

Have King's research into Jewish genetics and the surprising discovery of her own Jewish heritage challenged the way she thinks of herself? I asked her.

For the first time in our conversation, she paused, searching for the right words. "Clearly, I've been involved in genetics for decades ever since. I've gotten very involved in issues of ancestry." King paused. "If somebody asks me, literally, 'Are you Jewish?' I say, 'It depends on what you mean.' By Jewish law, I am not Jewish. My name, King, is not Jewish. I do not culturally identify as Jewish.

"I've thought from time to time, if I were not a secular person, would Judaism be closer to my belief set? 'Probably,' is the answer. As a geneticist, I think about that question in the same way that I think about the question of race. We define identity culturally and genetically. Most of us define our identity culturally in qualitative ways like language, religion, and nationality. The qualities have boundaries around them. But when we define our ancestry genetically, there are no boundaries, and it's a lot more complicated.

"It makes me wonder," she said. "Where do you start with a story of history?"

CHAPTER 13

SMART JEWS: JEWISH MOTHERS OR JEWISH GENES?

For Jews, history and identity are inseparable. While the headlines were shouting about the startling discovery of "Jewish genes" in the African Lemba, the back-of-the-Temple discussions were going in an entirely different direction: "Could black Africans really be Jews?" people whispered. Are the Lemba really blood brothers to Albert Einstein and Jerry Seinfeld? How did that happen?

"Here's what we would all like to know," wrote Christopher Hitchens, the *Vanity Fair* writer, who claims Jewish ancestry on his mother's side. "Did members of the Lemba minority furnish the majority of Southwest Africa's political revolutionaries, free-lance intellectuals, doctors, comedians, union-organizers, and lawyers? . . . [D]id the Lemba keep within their ghetto walls, muttering about the Second Temple and the tendency of their mutinous young people to marry Ovambo or Herero *shvartzes*?"* In other words, are the Lemba stereotypically Jewish? "I ask a serious question in a flippant way," he continued. "[O]ne has to ask whether Jews who think it kosher to 'think with the blood' are happy when

---

* A Yiddish word that literally means "black," but which is used derogatorily to denote an unsavory black person.

other groups do the same. The fact, of course, is that they (we) are not easy with this thought."

Many Jews see themselves as distinct from gentiles. Even among assimilated Jews and sometimes even in mixed marriages, there remains an acknowledgment of an "Us" and "Them" view of the world, a soft and silent but very real notion separating Jew from gentile. As the Harvard University professor Shaye Cohen writes, "Between Us and Them is a line, a boundary, drawn not in sand or stone but in the mind . . . [M]any rabbinic tests imagine that Jews are distinctive, identifiable, unassimilatable . . . The line is no less real for being imaginary, since both Us and Them agree that it exists."

As much as they might fiercely deny it for fear of provoking a backlash, Jews commonly partake in cultural or "racial" stereotyping when referring to themselves—they are members of "the tribe," bound by kinship and, regardless of geography, by blood and spiritual ties to biblical times. And while Orthodox Jews still bluntly claim God's unabashed favoritism, even secular Jews quietly indulge in self-congratulation about Jewish achievements.

Jews often draw pride, marked by self-mocking humor, at their reputation for being different and smart. They have coined a word, "meinstein," which translates into "my son, the genius." Jokes roll out. When does a Jewish fetus become a human? Answer: When it graduates from medical school. When Dwight Eisenhower met with Israeli Prime Minister David Ben-Gurion, the American president was said to have mused, "It is very hard to be president of 170 million people." Ben-Gurion reportedly shot back, "It's harder to be prime minister of 2 million prime ministers!" Many Jews with little or no knowledge of medieval Jewish history dish out folk tales of the Jewish marriage from heaven: the cleverest yeshiva boys in the shtetl were awarded the hands of the daughters of the richest moneylenders and merchants in the ghetto, guaranteeing that the smart genes would be passed along.

But pride in their history is tempered by the fear that publicly acknowledging Jewish achievements could feed the nasty stereotype of "Jewish domination"—of the World Bank, Wall Street, Holly-wood, academia, Washington think tanks, whatever. It threatens to resurrect Hitler-era notions of biological Jewishness. Somehow being Jewish or even "half Jewish" (rarely is one referred to as "half Christian," as if the imprint of "gentileness" is too faint to matter much) marks one, for better or worse, as distinct—and many people believe it's baked into the genes.

Suggestions that Jewish stereotypes are inborn may seem far-fetched and racist, but that belief has been embedded in the dialogue between Jews and gentiles for most of Western history. Well into the twentieth century, educated Jews and non-Jews alike unself-consciously used terms like the "Jewish question" and the "Jew-ish problem" when discussing the difficulties Jews had in escaping their history of separateness. But the plasticity of such stereotypes illustrates why the genetic default argument can be misleading and twisted to fit the ideology of the times.

Consider the age-old belief that Jews are less than stellar ath-letes. Recall the hilarious interchange in the 1980 comedy movie classic *Airplane!* between a flight attendant and a passenger. "Would you like something to read?" the woman asked. "Do you have something light?" he responds, to which she rejoins in a deadpan: "How about this short leaflet: 'Jewish Sports Legends'?" The scene never fails to touch off loud guffaws, even amongst Jews. For centuries, Jews have been caricatured as bookworms and physical weaklings, and to some degree that has been true: for most of their history, Jews were too impoverished to dally with athletics.

But history has a way of playing jokes on the unsuspecting. It might come as a surprise—a shock even—that Jewish athletes were once a major presence in American sports. For a brief few decades early in the twentieth century in the United States, Jewish immigrants poured into the urban centers of the East and Midwest, taking up various sports. Jews were a major presence in boxing, from world flyweight champion Abe "the little Hebrew" Attell at the turn of the century to the golden age of Jewish boxing in the 1920s and '30s, when Louis "Kid" Kaplan, Benny Bass, Izzy Schwartz, Barney Ross, and the heavyweight champion Max Baer, who had a Jewish grandfather and celebrated his partial Jewish ancestry by wearing a Star of David on his trunks, were stars. There were numerous prominent Jewish baseball players, from Hank Greenberg to Sandy Koufax. Chicago Bears Hall of Fame quarterback Sid Luckman helped usher in the modern version of football by mastering the forward pass.

The real shocker for many people is in basketball, where Jews were a dominating presence until World War II. Writers used to gush about the gaudy skills of the "natural" Jewish ballplayer. The superstars of the 1920s and '30s had names like "Inky" Lautman, Dutch Garfinkel, and Doc Lou Sugerman, and the top teams were the New York Celtics, led by Nat Holman and "Pretzel" Banks, the Cleveland Rosenblums, and the South Philadelphia Hebrew Association (known as the SPHAs or the Hebrews). In an incredible twenty-two-season stretch, the SPHAs played in eighteen semiprofessional championship series, losing only five.

"It was absolutely a way out of the ghetto," Dave Dabrow, a guard with the original SPHAs, told me before he passed away. The diminutive Dabrow spoke with the thick Yiddish accent so common among turn-of-the-century immigrant families. "It was where the young Jewish boy would never have been able to go to college if it wasn't for the amount of basketball playing and for the scholarship."

To the race-obsessed 1930s, a stereotypical form of "Jewish intelligence" seemed to make Jews particularly suited for basketball. "The reason, I suspect, that basketball appeals to the Hebrew with his Oriental background," wrote Paul Gallico, the sports editor of the *New York Daily News*, "is that the game places a premium on an alert, scheming mind, flashy trickiness, artful dodging and general smart aleckness." Jewish cleverness apparently helped make them "natural athletes," at least in basketball and boxing. After the war, as Jews began moving out of the inner cities, the traditional stereotype of Jews as brainy and indifferent to athletics (except owning teams, of course) began reasserting itself. Flash

forward to *Airplane!*, and the thought that Jews were natural athletes bordered on the absurd in the popular imagination.

It's axiomatic that Jews are by nature funny, but for most of
their history, they fought the stereotype of being dour, bespectacled
drudges who got their kicks from parsing obscure biblical conun-
drums. The Talmud is no competition for *Alan King's Great Jewish
Joke Book*. In medieval times, most Ashkenazi Jews, now known
as witty high achievers, spent their days in Ukrainian shtetls butch-
ering meat and running pawnshops. Most Jews lived among themselves in poverty and piety. Sure, there was the Yiddish conversation
style of challenging with a put-down or joke, but that was hardly
enough to pave the way for Fanny Brice and Groucho Marx.

If or when anthropologists discover that the Lemba are the
joke-cracking wizards of Africa, it would certainly not prove that
they share "funny genes" with Woody Allen, Roseanne, or Adam
Sandler. Certainly, Jewish communities around the world have been
amazingly tight knit for many centuries, but that's hardly evidence
that Jews are hardwired to be stand-up comics, clannish, naturally
gifted or hapless athletes, or have other stereotypical traits assumed
to be Jewish. After all, when compared to Jews, many African
Americans are as funny, Irish are as verbally facile, Indians are as
thrifty, and French are as persnickety.

So, what is in the genes? To what degree are there truths in
stereotypes?

The issue is contentiously debated. "Blank slate" theorists, in
ascendancy since after World War II, are resistant to ascribe group
traits to the genes. But an entire academic discipline blending genetics with sociology emerged in the 1960s to address this issue,
capped by the landmark publication in 1975 of Edward O. Wilson's *Sociobiology: The New Synthesis*. The distinguished Harvard
University biologist rigorously explained how Darwin's theory of
natural selection could extend to behavior and how it might work
in all species, from insects to people.

It's easy to understand why a parent might risk his or her life or
safety to protect a child. But a distant relative? Other members of a

community? Why do people volunteer to go to war for a country or an ideal? He offered a controversial evolutionary explanation for social behaviors, for why animal groups, often to the disadvantage of individuals within those groups, compete and cooperate, demonstrating kinship traits such as altruism, aggression, and nurturance. Sociobiology hit a raw nerve with liberal-minded sociologists and many geneticists, including Richard Lewontin and Stephen Jay Gould, who dismissed it as too speculative. It's a proxy for justifying the status quo, they wrote: if men are more likely to be in leadership roles, it's because they were programmed since their hunter-gatherer days to be that way.

Sociobiology, often called evolutionary psychology (although evolutionary psychology has a more cognitive focus), remains lacking in empirical evidence in the eyes of many scientists, but research and interest in the field has increased dramatically in recent years, spurred by new discoveries in population genetics. There is grudging acceptance that the founder effect and genetic bottlenecks that result in "ethnic" diseases may also contribute to group behavioral patterns. But which traits have a strong genetic component? Although most talk about inborn psychological or behavioral traits amongst Jews and other groups is plain rubbish or loose speculation, maybe not all of it is.

JEWISH IQ

The comedian and political commentator Al Franken often speaks about his education at Blake, an elite, mostly WASP prep school in suburban Minneapolis, which was once reputed to have restrictions on the admission of Jews. "They must have needed some Jews to get their SAT scores up," he jokes.

Franken is onto something. Some colleges acknowledge actively recruiting Jews in an effort to raise their rankings on SAT scores. According to the College Board, the average SAT for Jews, 1161, is the highest among any religious or ethnic group in the United

States and is considerably higher than the U.S. average, 1020.*
Eighty-seven percent of college-eligible Jews are currently enrolled
in college, versus 40 percent of everyone else. Jews make up 23 percent of the student bodies at the prestigious Ivy League universities
and 30 percent of the Ivy faculty.

This is hardly a new trend. By the 1920s, even while suffering
from widespread poverty, Jews were already overrepresented at the
Ivies despite a history of Jewish quotas. As one WASP graduate
of Harvard wrote in 1925 after an alumni gathering, "Naturally,
after 25 years, one expects to find many changes, but to find that
one's University had become so Hebrewized was a fearful shock.
There were Jews to the right of me, Jews to the left of me, in fact
they were so obviously everywhere that instead of leaving the Yard
with pleasant memories of the past I left with a feeling of utter dis-
gust of the present and grave doubts about the future of my Alma
Mater."

Once the intellectual doormats of Europe, Jews are now widely
thought to be the smartest people on earth. In the United States,
home to the world's largest expatriate Jewish community, Jews are
overrepresented in all forms of medicine, finance, academia, law,
television, radio, cinema, theater, journalism, and other prestigious
professions. Less than 0.25 percent of the world population is Jewish. Yet from a cold start in the early nineteenth century, by 1950,
the percentage of first-rank Jewish artists, musicians, and writers in
the world had soared to approximately 20 percent; in science, from
zero to almost 30 percent.

The social scientist Charles Murray analyzed the data to see if
they reflected merely the accomplishments and lack of prejudices
of the host country in which Jews lived, and he concluded that they
did not. He noted that from 1870 to 1950, Jews outperformed the
general population the most in France (19:1) and Germany (22:1),
with the United States (5:1), the most tolerant of countries to Jews

---

* Unitarians, who are a mixture of religious and ethnic groups, did score higher, at 1209.

during this time, near the bottom. Yuri Slezkine's *The Jewish Century* documents a similar trajectory of Jewish accomplishment in the former Soviet Union. Jews overcame considerable prejudice to rise to economic and cultural preeminence in almost every professional occupation. By 1939, one-quarter of university graduates and one-fifth of physicians and scientists were Jewish.

In every country with a significant Jewish population, the performance of Jews in high-achievement, high-paying careers has only increased in recent decades. How did that come to be? A laundry list of Jewish achievement doesn't tell you much about what Moses, Jesus, Baruch Spinoza, Maimonides, Albert Einstein, Sandy Koufax, and Jon Stewart (formerly Jon Stuart Leibowitz) might have in common. Is this record of unique achievement the consequence of nature or the Jewish focus on community and literacy? Jewish genes or Jewish moms?

High IQ certainly does not appear to have been embedded in the genes of the early Israelites. The early Jewish population during the first millennium BCE was at times as large as or larger than the citizenry of ancient Rome and Athens, but its intellectual contributions were not comparable. While Jews wrote the Bible over many centuries, an enduring although singular contribution, the Greco-Roman world revolutionized art, science, and literature.

The first historical notation of Jewish high achievement shows up in Babylon, the Jewish diaspora outpost, where Jews were amply represented among the societal elite, despite their second-class citizenship. Natural selection does not need to be invoked as an explanation. According to Jewish history, the accomplished and educated Jews were exported to Babylon in 587 BCE, while the poor, ill-educated peasants remained in Judea, forming the core of the surviving Jewish population. The Judeans developed a reputation not as intellectuals but as religious zealots and rebels, at least in the eyes of Christian chroniclers like Edward Gibbon, author of *The Decline and Fall of the Roman Empire*.

There are other examples of Jewish secular achievement from biblical times until the nineteenth century, but not many. There is no evidence that they played a major role as merchants during the rise and fall of the Roman Empire, a major Jewish outpost for nearly a millennium. Even in the golden age of Sephardic Jewry in Spain, except for Maimonides, and until the 1800s, except for Spinoza, Jews were almost absent from the annals of great science and art. Then, like the flip of a light switch, in a matter of decades, the Ashkenazi Jews of Europe had become known as intellectual thoroughbreds.

"It seemed as if a huge reservoir of Jewish talent, hitherto dammed up behind the wall of Talmudic learning, were suddenly released to spill over into all fields of Gentile cultural activity," wrote the historian Raphael Patai. The Christian Enlightenment and its Jewish counterpart, the Haskalah, opened Jewish minds, and the European emancipation movement opened cultural doors. Once derided for their backwardness, Ashkenazi Jews began to be described with mixed awe and concern about their financial acumen and cultured scholarliness.

"The Jews presented . . . in our day, in proportion to their numbers, a far larger list of men of genius and learning than could be exhibited by any Gentile country," commented Lord Ashley, the future Seventh Earl of Shaftesbury, at an 1847 debate in the House of Commons on whether to allow non-Christians to properly swear the oath of membership. "Music, poetry, medicine, astronomy, occupied their attention, and in all they were more than a match for their competitors."

Not all Christian intellectuals considered Jewish accomplishments an indication of high intelligence or culture. The great German composer Richard Wagner captured the sentiment of many Europeans irked by Jewish proclamations of their biblical uniqueness and the upstart success of these historical vagabonds. In his widely read 1850 essay "Jewishness in Music," he aired his irritation over the fact that the operas by rival composer Giacomo Meyerbeer, a French Jew, were more popular than his. "Our whole European art and civilization . . . have remained to the Jew a foreign tongue," Wagner wrote. "In this Speech, this Art, the Jew can only mimic and mock—not truly make a poem of his words, an art-work of his doings." (Hitler would later adopt the *völkische* view that Jews were nomadic cultural parasites: "The Jewish people, de-spite all apparent intellectual qualities, is without any true culture, and especially without any culture of its own," he wrote in *Mein Kampf* (*My Struggle*), his Nazi manifesto published in 1925. "For what sham culture the Jew today possesses is the property of other peoples, and for the most part it is ruined in his hands.")

Most educated Ashkenazi Jews of this era proudly if quietly em-braced the notion of a Jewish "race" of smart high achievers. Most scientists caught the fever of race science, including the noted Jew-ish anthropologists Joseph Jacobs, Samuel Weissenberg, and Ignaz Zollschan. Jacobs theorized that intellect and physical strength are inversely linked, like a teeter-totter. "If [the Jews] had been forced by persecution to become mainly blacksmiths, one would not have been surprised to find their biceps larger than those of other folk," he wrote. "[S]imilarly, as they have been forced to live by their exercise of their brains, one should not be surprised to find the cubic capacity of their skulls larger than that of their neighbors." Intelligence was God's gift for shortchanging them physically. With towering role models such as Freud and Einstein and a well-known commitment to education, the belief in the intellectual über-Jew was firmly established in Europe by the early twentieth century.

This was interesting speculation in this pregenetic era, but there was not much empirical evidence to back it up. Scientists relied upon surveys, such as one in London in 1900 that found that the mostly poor immigrant Jewish population disproportionately won academic prizes and scholarships. But that was far from convincing data.

That changed with the development of IQ tests. Developed by the French psychologist Alfred Binet in 1904 as a way to evaluate learning disorders in children, the test was modified by the Stanford University psychologist Lewis Terman as a way to measure cogni-tive ability—what scientists came to call general intelligence, or *g*. Based on a scale in which those scoring 90 to 110 were deemed av-erage, in 1916, Terman developed the Stanford-Binet Scale, which became known as the Intelligence Quotient, or IQ, test. What did the early IQ tests show about Jews? In one of Terman's most famous studies, of California children in the early 1920s, 10.5 percent of those scoring 135 or higher were Jewish. Considering many Jews of this era hid their religious affiliation, Terman was convinced the percentage was far higher. A survey of scores in London in the 1920s at three schools of students from different economic classes showed that Jews at each school scored nearly fifteen IQ points— one standard deviation*—higher than non-Jews, about what most tests show today.

Critics of IQ tests ridiculed Terman's findings, pointing to an earlier 1913 study by Henry Goddard, a prominent early-twentieth-century psychologist and eugenicist. Goddard purportedly found that 83 percent of Jews (along with 80 percent of Hungarians, 79 percent of Italians, and 87 percent of Russians) spilling off ships at Ellis Island tested as "feebleminded." To skeptics, Goddard's claims demonstrated the fundamental absurdity of IQ tests. After all, they said, with no hint of irony, what could be more ridiculous than to believe in a test that showed Jews were morons?

But that's a distortion of what Goddard reported. The mischaracterization appears to trace to a 1974 book, *The Science and Politics of IQ*, by Leon Kamin, an American psychologist who has campaigned for decades that the heritability of IQ could be as little as zero. In fact, Goddard was in the process of translating the Binet test into English and standardizing and normalizing the first IQ tests when he undertook his study. He specifically focused on the "feebleminded" and the "defective," terms commonly used by scientists of that era to describe people with low intelligence.

---

* Standard deviation is a statistical measure of the spread of data. One conceptual way to think about the standard deviation is that it measures how spread out the bell curve is. These parameters give an easy way to summarize data: 68 percent of the values are within one standard deviation of the mean; 95 percent of the values are within two standard deviations of the mean; more than 99 percent of the values are within three standard deviations of the mean. An average measured IQ of any population of approximately 115 would be considered extraordinary.

His rudimentary IQ test given to some 152 immigrants was able to identify those suspected of being retarded, although even he acknowledged that the percentage of those who fell into that category was way too high, as the IQ tests were far too crude at that point to be reliable.

It's now widely recognized that those early IQ tests were flawed for any number of reasons. During the 1920s and '30s, they were often directed at immigrants fresh off the boat from Europe, arriving in a country that was turning increasingly hostile toward what many considered an alien invasion. And the tests themselves were hopelessly biased. Although they were supposed to measure innate intelligence, they were filled with questions that only longtime U.S. citizens could be expected to know, including questions about American sports, such as baseball ("Who is Christy Mathewson?"), and consumer products ("What is Crisco?").

By the 1940s and '50s, IQ tests had become far more refined and less culturally skewed. It's problematic to draw too much from selected studies, but over the past fifty years, Ashkenazi Jews have consistently tested well above the norm in almost every IQ study. Some of the findings were startling. In one 1954 review of New York City's standardized IQ results, twenty-eight children were found to have scored 170 or higher—twenty-four of them were Jewish. Looking at worldwide population groups, Jews appear at the top of the IQ charts, followed by East Asians, whites, Arabs, other Asians, Indians, blacks, and Australian Aborigines. Studies have set average Ashkenazi g at anywhere from 107 to 117, which would rank them as the highest tested population in the world, as much as a full standard deviation above the general European average, of about 100.

Jews not only rank higher in average IQ, the structure of their intelligence is different than that of other groups. "Considered as a group, [Ashkenazi Jews] tend to excel in some cognitive domains— for example, verbal and numerical ability—but not in others, as witness their unexceptional performance on certain types of spatial or perceptual problems," wrote Miles Storfer in his well-regarded

1991 review of the nature-nurture debate on intelligence. Jews test especially high on verbal ability—one estimate puts this component at an average of 125. But in a few tests in which visuospatial ability has been measured, they test lower than the European average. Non-Ashkenazi Jews—Sephardim and Oriental Jews who have undergone more intermixing—do not have higher average IQ scores, nor are they more likely to be in high-achieving jobs.

While the IQ differences between Jews and northern Europeans may not seem large, in practical terms, they could result in a dramatically higher proportion of Jews with very high IQs. Ashkenazi Jews are concentrated at the smart (right) end of the distribution curve, where geniuses reside. A metastudy of numerous IQ tests found that one-fourth of the white Americans with IQs above 145 are Ashkenazim. Another study found that only four per thousand northern Europeans have an IQ over 140, as compared to twenty-three per thousand Ashkenazim. In other words, a Jew is six times more likely than other whites to be considered a genius.

IQ tests are significant because they are among the most reliable predictors of success in university studies and in many careers. Organizations from the U.S. military to corporate America to the National Football League use intelligence tests to assess a person's chance for success at a specific task or job. It's hardly surprising, therefore, that Ashkenazi Jews not only score higher on IQ tests, but they also are disproportionately represented as high-intellect achievers. Citing just a few examples, the first Nobel Prize awarded to a Jew was given in 1905, four years after the prizes were initiated. By 1950, 14 percent had gone to Jews. Over the next half century, 29 percent of the winners were Jewish or of substantial Jewish ancestry—a success rate twelve times higher than would be expected based on population figures alone. Jews have captured 40 percent of the prizes in economics, 28 percent in physiology and medicine, and 26 percent in physics. Jews and "half Jews" account for 20 to 30 percent of Nobel Prize winners and 36 percent of all U.S. winners. They've won more than one-quarter of the Westinghouse Science prizes, the ACM Turing Awards (considered the

Nobel Prize of the computing world), and the Fields Medal (the top mathematics award). Fifty-four percent of the world's chess champions have recent Jewish ancestry. Although we should be cautious in interpreting the causes of this success, it's impressive.

BRAINS AND IQ

We can agree that Jews outperform other groups in many fields that place a premium on high IQ. But why? Two questions come to mind. Is intelligence innate? And if so, is there evidence that Jews, Asians, or any ethnic group or population is smarter, on average, than any other? These ticklish questions hint at resurrecting race science, so it's understandable that asking them makes us nervous. But it's these kinds of questions that we as a society must face in coming years, as the fruits of human biodiversity research ripen.

Simply said, genes circumscribe human possibility. That's relatively uncontroversial when we talk about body types. Along with cultural and environmental factors, genetically based physical and physiological differences—from muscle fiber type, to lung capacity, to limb-to-trunk ratios, to metabolic efficiency—go a significant way toward explaining why Eurasian whites dominate weight-lifting and field events, East Africans disproportionately win the world's top distance races, and the best sprinters trace their primary ancestry to central West Africa.

But does that hold true for the brain, the most complex of body parts? Which has more of an influence on intelligence—DNA or the environment? It would be nice if intensive education or even switching parents could significantly raise IQ, but the evidence for that is scant at best. Studies of identical twins raised in radically different environments show very little sustained differences in aptitude. IQ-focused educational programs, such as President Bush's "No Child Left Behind," may raise test scores by as much as eight points, but the effects are temporary, largely the result of more efficient test taking, and soon fade away.

Even discussing this subject puts us in dangerous territory. By definition, any measure of intelligence is a ranking system. While everyone can embrace the concept of being unique, none of us would be happy to be considered inferior, as low IQ implies. Historically, ruling racial or ethnic groups conceived systems that ranked their members at the top of the virtue and intelligence status pole. That list has included biblical Israelites (with Canaanites and other "pagans" considered deficient) and subsequent "great civilizations," from the Greeks and Romans (the barbarians, Christians, and Jews were inferior), ancient Chinese (the rest of the non-Asian world was deficient), the Japanese (whites were slackers), or "Aryan German Christians" (Jews, Gypsies, and blacks, among others, failed the racial purity test). The scientific worldview, when it crystallized in Europe in the nineteenth century, ranked whites, men in particular, at the top of a scale that had those with "colored" skin at the bottom.

This poisonous history has made it almost impossible to have a reasoned public discussion about the causes of human differences, especially intelligence. The classic way of discussing this issue—is it nature or nurture?—simply doesn't capture the complexity of human development. Ideology and not science often divides those inclined to see the world through the deterministic prism of genes from those who embrace the flexibility of the environment. Growing up poor or malnourished can swamp most inborn advantages. There is also a feedback loop between nature and nurture. Many genetic differences can be ascribed to cultural and environmental forces: geographic isolation, natural disasters, education, prenatal health, social taboos (such as not marrying outside one's faith or tribe), and even one's diet can determine who survives to pass on his or her genes. Many gene functions are expressed, or activated, by environmental triggers. It's still not understood what factors flip the on and off switch in proteins that tell cells how to develop. Because the protein functions in our brain can be altered by experience, it's certain that genes alone cannot explain the complex workings of the brain. What we do know is that DNA matters—a lot.

Because the brain is part of the body, it is surely as susceptible to evolutionary forces as any other body part. Any parent with children can attest to the fact that intelligence, temperament, and behavior are to some degree out of their hands. The authors of the Bible certainly understood that. Rebecca and Isaac struggled mightily to raise their fraternal twin sons. "Two nations are in your womb," the Lord declared to Rebecca in Genesis 25:23. "Two separate peoples shall issue from our body." Esau was born red and hairy; Jacob was smooth-skinned. While Esau grew up to be a fearless hunter, Jacob was sensitive, mild-mannered—and Machiavellian. The differences were hardwired.

Modern twin studies show similar genetic differences. While the intelligence, personalities, behavior, and even career choices of identical twins are astonishingly alike, siblings of the same sex, including fraternal twins, raised in the same home are often strikingly different in all aspects, including mental capacity. Identical twins raised separately following adoption show an astounding correlation of 0.72 for general intelligence, *g*, which is reflected in IQ scores. Even skeptical social scientists now acknowledge that genetics is the major influencer of intelligence. An ideologically and racially diverse blue-ribbon scientific panel commissioned by the American Psychological Association (APA), hardly a hotbed of radicalism, has concluded that IQ tests measure *g*, different aspects of intelligence are interconnected, and IQ can predict achievement in everything from college success to income.

It used to be that scientists believed intelligence was highly and straightforwardly correlated with brain size; now they think that intelligence, like real estate values, depends on "location, location, location." Consider Albert Einstein's brain, the morphological metaphor for genius. When Einstein died in 1955 at age seventy-six, Thomas Harvey, the pathologist who performed the autopsy, cut his brain into pieces and preserved it in formaldehyde. The overall size of Einstein's brain was unremarkable. It actually weighed a third of a pound less than the 3-pound average of adult males. Size itself didn't matter. However, the structure of it was unique. It had greater-than-normal numbers of glial cells, which are needed to nourish high-performance neurons and are believed to help the hippocampus and cerebellum work more efficiently.

Harvey turned the brain over for study to Sandra Witelson, a neuroscientist at McMaster University in Hamilton, Ontario, where it and more than one hundred others are locked away in a fortresslike walk-in brain bank refrigerator. She compared measurements and photographs of his brain with the preserved brains of men and women known to be of normal intelligence when they died. Einstein's brain fell in the range of normal, except for the portion in the middle of the brain known as the inferior parietal lobes, which influence mathematical and music ability and help process visual images. Einstein's parietal lobes were 15 percent wider, and the neurons in this area were packed more tightly than in other people. She also found that the groove known as the sulcus, which normally runs from the front of the brain to the back, did not extend all the way in Einstein's case. She believes this fissure, which appears very early in life, probably did not shrink because Einstein used his brain a lot but was likely always absent.

"We don't know if every brilliant physicist and mathematician will have this same anatomy," Witelson said in 1999. "It fits and it makes a compelling story, but it requires further proof."

Proof or at least more compelling evidence arrives every year. Witelson herself has released a study that found a high correlation between the size of certain sections of the brain and intelligence, particularly verbal ability. On average, what's called the packing density of the neurons in the language region of the temporal lobe in the adult female brain is 12 percent greater than in the adult male brain. The major pathways that connect the left and right cerebral hemispheres, known as the corpus callosum, are also bigger and more developed in females. (Magnetic resonance imaging tests, however, have shown little or no difference in the corpus callosum of men and women.) All of these factors could help explain women's greater communications skills (and maybe why girls are so enamored with text messaging!).

Gender differences are almost as taboo a subject as race, but Witelson has not shied away from the scientific evidence. "What is astonishing to me is that it is so obvious that there are sex differences in the brain and these are likely to be translated into some cognitive differences, because the brain helps us think and feel and move and act," she has said. "Yet there is a large segment of the population that wants to pretend this is not true." She has been critical of those who targeted the former Harvard University president Lawrence Summers for suggesting that innate differences might help account for why there are more men than women on science faculties. "If we're going to try to understand the disparity between the number of women and men in different professions, and this would go for positions way beyond just academia, we have to put all the factors on the table. It's clear societal influences are relevant, but that doesn't preclude the possibility that there are also contributing factors from nature."

Brain maps confirm links between genes and intelligence and their location. Cognitive ability has been found to be linked to the gray matter in the frontal lobe, the lateral prefrontal cortex, which is considered the seat of intelligence and the part of the brain influenced most by the genetic makeup of our parents. When a person uses certain definable skills—for example, those that require spatial reasoning, which is a major component of high IQ—the gray matter lights up. Overall IQ as well as particular mental strengths and weaknesses, such as in verbal or mathematical reasoning, correlates with neuron density found in the gray matter. It's believed that genes interacting with environmental stimuli determine how key sections of the brain develop. The heritability of some brain functions tops out at an astounding 90 to 95 percent.

"This may be why one person is quite good at mathematics and not so good at spelling, and another person, with the same IQ, has the opposite pattern of abilities," said Richard Haier, a professor of psychology and a leading MRI researcher at the University of California at Irvine. The Yale University psychologist Jeremy Gray and Paul Thompson, a neurologist and neuroimaging specialist at the University of California, Los Angeles, recently reviewed all the relevant studies. "In our view—which is shared by most investigators," they concluded, "the data unambiguously indicate a neurobiological basis for intelligence, particularly for reasoning and novel problem-solving (which strongly predicts psychometric *g*)."

Skeptical social scientists often say, "Show me the intelligence genes," as if any one particular human trait, let alone the functioning of nature's most complex organism, should be able to be reduced to one or even a few genes. Geneticists scoff at such simplisms. It's unlikely that intelligence is totally fixed at birth by a certain set of genes. Rather, the capacity for intelligence is probably a probability. Children of smart parents may well inherit certain genes that incline them to interact with their environment in a very stimulating way. Considering the incredible complexity of the brain, the mosaic that makes up reasoning ability will undoubtedly be affected by the interaction of various genes, perhaps hundreds, expressed through the prism of the environment and experience.

The exact relationship between DNA, the brain, and intelligence is still cloudy, but genes influencing cognitive ability do exist, and science is slowly identifying them. For example, several genes for schizophrenia are associated with lower cognitive abilities in both schizophrenic and pediatric populations. Based upon the finding that there is a common biochemical mechanism at the root of learning and memory, researchers have developed a way to switch on and off in mice the gene NR2B, which controls the brain's ability to associate one event with another, the core feature of learning. This receptor, found in the forebrain and the hippocampus, is almost identical to the receptor in humans, although it's not clear that it works in the same way.

The most intriguing new research involves genes that might have been the trip wires for modern civilization. How did humans get to be so smart? Why are we living in cities instead of caves?

The study of human intelligence, dotted as it is with political land mines, is dicey stuff. But a University of Chicago research team led by Bruce Lahn has made some headway in this controversial field. Using statistical analysis, he traced the microcephalin gene, which regulates brain size, to more than 1 million years ago. It was carried by an archaic human population and made its way into the modern human gene pool, possibly through one tryst between the two related species, approximately 37,000 years ago. That's about the time when Neanderthals, who had larger brains than modern humans, were fighting a losing battle against newly arriving Africans.

Although preliminary research on the tiny bit of DNA recovered from a Neanderthal fossil so far indicates no clear evidence of interbreeding, a forty-thousand-year-old skull found in a Romanian cave shows traits of both modern humans and Neanderthals, throwing the scant DNA data into doubt. It's possible that a mystery offspring had a hybridized brain and proved wildly successful. Intriguingly, the arrival of the microcephalin allele coincides with the emergence in Europe of art, symbolism, and the anatomical evidence for the ability to speak. It must have provided a strong survival benefit for it to have spread to a majority of the human population while the Neanderthals died out.

Lahn's team also found variants of another brain gene, ASPM, dating its appearance to 5,800 years ago. That's about when cities and written languages first flourished in the Middle East. Coincidence? Possibly. Although it's not clear how these two genes work, they both appear to stimulate the growth of neurons in the cerebral cortex, resulting in the development of a neurologically richer brain. They could also have any number of other effects, including changing prenatal head size to make childbirth less risky or improving energy efficiency in the brain. Whatever the effects, the progeny of humans with these brain mutations overwhelmed their dim-witted brothers and sisters. It's positive selection.

The findings have a racial wrinkle. The microcephalin gene shows up in as many as 70 percent of Europeans and East Asians but is much less common in sub-Saharan Africans, where 25 percent or less of the population carries it. ASPM is carried by 44 percent of Caucasians from the Middle East and Europe, where it is believed the mutation first appeared, but is rarer in East Asians and is almost nonexistent in black Africans. Is it just coincidence that the northern and eastern branches of modern humans quickly moved from drawing glorious cave paintings to developing modern agriculture and founding villages while the southern humans who then made their way to Australia and the Pacific remained locked in a far more primitive culture?

This brain research threw another log on the fire of the intelligence and race controversy. Francis Collins, head of the federal genome project, criticized linking brain genes with innate intelligence. Spencer Wells reiterated his refusal to study the brain in his Genographic Project, choosing to stay on the safe ground of physical differences. Social scientists were apoplectic, invoking the familiar whipping boy, *The Bell Curve*, by Richard J. Hernstein and Charles Murray, and more. "He [Lahn] is doing damage to the whole field of genetics," said Pilar Ossorio, a professor of law and medical ethics at the University of Wisconsin. The vociferous response is believed to be behind a decision by the University of Chicago to abandon a patent application to cover a DNA-based intelligence test based on Lahn's work.

Their caution, if not their invective, has merit, considering the complicated nature of intelligence and the extreme difficulty of measuring genetic effects on such an abstruse concept. Although a scientific consensus has developed around the notion that intelligence has an innate biological component, differences between population groups remain a contentious issue. The fact that individual intelligence is highly heritable does not necessarily mean that group differences are. In fact, when scientists crunched Lahn's data to see if people who carried these mutations had higher IQs or even larger brains, they found no linkage. The possible evolutionary advantage of the mutations may show up in other adaptive ways that led to advances in civilization, such as the ability to mentally focus over an extended period of time, which is not measured in standard

IQ tests. Already, a relationship has been found between these alleles and verbal intelligence, such as the ability to process written and oral texts.

Lahn, who is of Chinese ancestry, has been stunned by the backlash. The details of his "civilization threshold" theory are beside the point, he said; the mutations almost certainly have proved critical in brain evolution. He emphasized that it would be a gross distortion to interpret these alleles as proof that one ethnic group might be more or less "evolved" than another. Brain development almost certainly results from a multitude of genes likely to be found in each population group. But Lahn fears that what he calls the intellectual "police" in the United States might make such questions impossible to pursue. "Society will have to grapple with some very difficult facts" as scientific data accumulate.

ASHKENAZI IQ

"I think that Bruce doesn't understand political correctness," said Henry Harpending, an evolutionary anthropologist at the University of Utah. Harpending should know. With the recent publication of his coauthored treatise on Jewish intelligence, he's a target as well.

For two years, beginning in 2003, Harpending and Gregory Cochran, a physicist turned genetic theorist, honed a controversial theory that embraced, with notable scientific flourishes, the Jewish folk explanation for high IQ: Ashkenazi Jews are smart because they are born that way. Eastern and central European Jews evolved their higher intelligence during the Middle Ages because they were forced to work mainly in occupations that required greater cognitive ability. It's a by-product of anti-Jewish discrimination, Jewish separateness, and the historical Jewish commitment to education, all of which influenced gene evolution.

This theory had been put forth before. But Cochran and Harpending riveted the attention of the chattering classes by also seeking to resolve another scientific mystery: the odd cluster of brain and nervous system disorders that stubbornly persist among Jews. Maybe centuries of moneylending and dissecting the inscrutable Talmud had paid off, but at a high price. Cochran, who developed the theory, dubbed it "overclocking"—computerspeak for eking out extra performance. The problem with overclocking, Cochran has said, is that "sometimes you get away with it, sometimes you don't"—some geeks run their hard drives faster than they were designed to, which can cause crashes or breakdowns. Could this be what has happened to Ashkenazi Jews?

The paper made the rounds on highbrow Internet discussion lists, where it was attacked, extolled, and refined. After recruiting the University of Utah social scientist Jason Hardy as another co-author, the authors circulated it to various journals. Considering its subject matter, let alone its speculative science, it is not surprising that the paper got no takers at first. Finally, in summer 2005, the *Journal of Biosocial Science*, published by Cambridge University Press, posted "Natural History of Ashkenazi Intelligence" online and agreed to run it in its journal in 2006. The article became an overnight sensation, touching off a debate that ricocheted through the blogosphere, was discussed by Jews with quiet pride and genuine concern, and eventually broke into the mainstream with respectful reviews in the *Economist*, *New Scientist*, the *New York Times*, and even *New York* magazine, which featured on its cover a grinning Larry David (huh?) as the emblematic "smart Jew."

Did the authors offer definitive proof for their speculative theory? No. Their case could be tested using twin adoption studies that measure IQ differences of siblings raised by Ashkenazi and gentile parents but no such studies exist. The authors were forced to rely on intriguing circumstantial evidence. One common thread emerged in the unusual cluster of metabolic and neurological disorders known as lysosomal storage diseases (LSDs). Lysosomes are the cells' garbage disposal system, containing enzymes that digest worn-out cells and food particles and engulf viruses and bacteria. When lysosomes don't work right, waste products accumulate in the cells. These defects can disrupt the way information passes through the sphingolipids, the insulating shell of nerve cells. There are more than forty LSDs. The four common recessive ones that afflict many Ashkenazim—Tay-Sachs, Niemann-Pick, Gaucher, and mucolipidosis type IV—cause a host of issues, including spleen and bone problems, mild retardation, and even death.

Harmful mutations usually disappear because people who carry mutant genes often die at an early age or have difficulty in finding mates. Why weren't these deadly disease genes passed out of the gene pool, eradicated by natural selection, as were so many past scourges? There are exceptions to the grinding work of nature, usually when a mutation offers some survival benefit. That may be why the LSDs have persisted. It turns out that survivors who suffer from these diseases share another similarity besides their Ashkenazi Jewish ancestry—they are often of unusually high intelligence. In other words, these mutations may offer real but costly benefits for survivors.

There are other examples of evolutionary trade-offs. Consider the sickle cell gene. If you carry one copy of the mutation, you won't get the disease and you're less susceptible to malaria. But if you inherit one mutated-gene variation from each of your parents, you can end up with debilitating anemia, beset by blood clots, fatigue, and retarded growth. The potentially harmful effects of the sickle cell mutation are balanced against its malaria-protective abilities, which for thousands of years kept the gene from disappearing in Africans and other populations. Some researchers believe that blacks may suffer disproportionately from another disease, type 2 diabetes, which researchers speculate is the consequence of a "thrifty gene" that enabled the African ancestors to use food energy more efficiently when food was scarce. In the modern world, the gene that once aided in survival may instead make the gene carrier more susceptible to developing diabetes.

Could the LSD mutations work the same way? Perhaps two copies can leave the brain wasted; one copy and you're smart?

That's exactly what Cochran and his team suspect shaped

Jewish intelligence. Although they have not yet proved that any disease genes actually affect intelligence, they have a theory on how the process might work. They speculate that genes that stimulate the extra growth and branching connecting nerve cells together might promote intelligence. It's obviously not definitive, but that's what happened in a 1995 study on rats afflicted with Gaucher. If there were two copies of the gene—recessive mutations donated by the mother and father—a chemical in the brain built up, and the branches on the neurons grew wildly, leading to a debilitating brain disease. However, if there was only a single copy of the gene, the chemical buildup led to more intense but relatively controlled growth in the brain. That's what appeared to happen to the rats with a single copy of the Gaucher mutation. Could it work that way in humans?

Gaucher is the most common disease found disproportionately in Jews, with the mutation showing up in 6 percent of Ashkenazim. There is at least circumstantial evidence linking Gaucher to high IQ. At Cochran's request, the Gaucher Clinic at Jerusalem's Shaare Zedek Medical Centre furnished him with a list of the occupations of more than 250 adult patients, essentially all the adult Gaucher sufferers in the country. "Many of my patients are well-known intellectuals," said Ari Zimran, director of the clinic. Fifteen percent of the patients were engineers or scientists, versus an estimated 2.25 percent in the Israeli Ashkenazi working-age population, and there were twenty times as many physicists as might otherwise be expected. That's not slam-dunk science, but it's intriguing.

There are at least two other non-LSD disorders unusually common in Jews that show indirect evidence that they affect intelligence: congenital adrenal hyperplasia (CAH) and torsion dystonia. CAH results from a disruption in the production of the adrenal steroid hormone during fetal development. It can lead to an overproduction of sex steroids, which can result in infant girls being born with more malelike behavior and physical characteristics. Girls with CAH test more like males on IQ tests, scoring significantly higher in abstract reasoning, a skill that many scientists believe may be shaped in part by sex hormones.

Torsion dystonia is a progressive muscular disorder that can leave the body twisted into a pretzel. There are anecdotal accounts of victims of the disease with high IQ; one study found the average IQ for victims was 121. Sharon Drew Morgan of Austin, Texas, tells the story of her son, who talked at nine months, said a hundred words before he said "mama" or "dada," spoke in sentences at twelve months, and read at three. At five years old, he was tested to have an IQ of 145, in the genius range. Years later, she says, "I woke up to a limping nine-year-old." Her son's left foot would not fully extend, and he walked on his heel. The problem spread to his right leg and his hands, and by the time he was ten, he was crippled, a victim of generalized torsion dystonia. Scientists have long been aware of the link between torsion dystonia and intelligence. Two 1970 studies encompassing more than six hundred cases found victims with the recessive form of the disease—almost all of them Ashkenazi Jews—had measurably "superior intelligence."

INTELLIGENCE AND DISEASE

Could the Cochran theory explain why so many "Jewish diseases" have not been erased by natural selection? Because of cultural factors honed over many centuries, do Jews choose mates with higher intelligence and risk the possibility of having children with debilitating diseases? By some mysterious, evolution-driven psychic calibration, do Jews value brains over health? Is positive selection at work?

It's safe to say that most scientists are reluctant to embrace positive selection as an explanation of Jewish intelligence, at least for now. A decade ago, Jared Diamond of UCLA, University of Utah geneticist Lynn Jorde, and other scientists speculated that in the case of Tay-Sachs, the mutation might have offered protection against "ghetto diseases," like tuberculosis, smallpox, and bubonic plague, which ravaged the poor populations of Europe during the Middle Ages. Yet, as Stanford University's Neil Risch has pointed out, French Canadians are prone to a version of Tay-Sachs that originated in a different founder, but which occurs at almost the same frequency as among Ashkenazim, but "I've never heard anybody arguing that Tay-Sachs provides *them* an advantage against tuberculosis in crowded ghettos!" (Risch's exclamation aside, French Canadians, like Jews, are descended from a small founder population—thousands of French landed in Quebec some three hundred to four hundred years ago, many of whom, like Jews, lived for centuries in poor, insular communities.) Most ghetto diseases are now largely under control and today rarely affect Jews (or French Canadians), and yet the disease gene for Tay-Sachs in Jews persists.

Cochran agrees with Risch on this point, but that does not undermine his intelligence theory. "If [the LSDs are positively selected to protect against diseases], many other groups, including neighboring Poles and Russians, who suffered from similar squalid conditions, would have high frequencies of these or similar mutations," Cochran said. In fact, there is little sign of these diseases in Christian Europeans. "But that doesn't mean there was no positive selection at work." These double-edged genes were selected to perpetuate intelligence, not to protect against disease, Cochran suspected.

Risch countered with an alternative explanation—random (or in this case, bad) luck, also known as genetic drift, combined with the bottleneck or founder effect. (That's when a mutation appears by chance in a geographically or culturally isolated population, like Jews, and then spreads.) Risch dated torsion dystonia to a lone Lithuanian Jew who lived twelve to thirteen generations ago, which roughly coincided with the devastating Thirty Years' War and the Chmielnicki uprising in the first half of the seventeenth century. He believed that bottleneck also explains the spread of the four LSD diseases prevalent among Jews.

To test his thesis, Risch rounded up LSD sufferers identified for him by Rabbi Ekstein, known for his pioneering work in containing Tay-Sachs, and compared their genes with Ashkenazim with other common Jewish diseases. He found the diseases all appeared to have originated as single mutations. "These observations provide compelling support for random genetic drift (chance founder effects, one 11 centuries ago that affected all Ashkenazim and another 5 centuries ago that affected Lithuanians), rather than selection, as the primary determinant of disease mutations in the Ashkenazi population," Risch and his coauthor Hua Tang concluded. "It seems implausible that all these disease mutations have undergone selective advantage unique to one population."

Risch's attack on the positive selection thesis has its own fundamental weakness, however. Genetic drift can account for the origins of these mutations but does not easily explain why they have survived the grueling gauntlet of natural selection. "I'm not convinced that the bottleneck hypothesis is the sole explanation," said Joel Zlotogora, the director of the Department of Community Genetics for the Israeli government. "I think it's very difficult to explain [the persistence of] these diseases just by chance," he told me.

With understandable astonishment, Zlotogora and fellow Israeli geneticist Gideon Bach of Jerusalem's Hadassah-Hebrew University Hospital observed that the four LSDs cluster in one enzyme pathway out of the thousands of known biochemical paths. They conclude that it's improbable that so many common mutations would randomly originate in exactly the same place, as Risch's explanation would have it. They also note a clear distinction between the brain and neurological disorders linked to high IQ and many other Ashkenazi diseases, which they agree are due to chance. In many Jewish diseases, only one mutation is prevalent—which is not the case in the four LSDs. There are three common Tay-Sachs and Niemann-Pick mutations, for example. Could they have appeared independently in diverse locations at different times purely by chance—and then not been selected out of the gene pool, as harmful genes usually would be?

"People could say it's chance, I suppose," said Cochran, "in the same way it's chance that [Jews represent] 27 percent of all those guys who go to Stockholm every year."

Cochran is also convinced positive selection may be at work in the other major cluster of Ashkenazi mutations that affect the same enzyme pathways—the DNA repair cluster, including the breast cancer genes, Fanconi anemia (stunted growth and bone marrow, heart, kidney, and spinal defects), and Bloom syndrome (shortness, a narrow face, and a high-pitched voice). They reason that the four LSDs and these related mutations might all unleash neural growth in a way that favors cognition. They may be onto something. The BRCA1 gene, which suppresses tumors when not mutated, appears to have been positively selected in humans but not chimps after they diverged from each other. Why would a deadly cancer gene be positively selected? Geneticists at the University of Chicago believe the BRCA1 gene may contribute to brain development. In other words, killer mutated versions of BRCA1 may have stubbornly persisted in the Jewish gene pool because the likelihood of this allele promoting high intelligence may override the lesser possibility of its causing cancer.

As speculative as the Cochran hypothesis may seem to critics, Jewish history is on its side. Jared Diamond floated a similar but more simplistic version of this theory in 1994. The mutated lysosomes might have been positively selected "in Jews for the intelligence putatively required to survive recurrent persecution, and also to make a living by commerce, because Jews were barred from the agricultural jobs available to the non-Jewish people," he wrote.

As we've read, from the ninth century onward, most European Jews were periodically locked in ghetto communities, their interactions and indeed their survival often dependent upon the unique skills they offered to their Christian hosts. Many Christians were forbidden from money-related professions, such as banking and tax collecting. Eastern European Jews often survived by usury and by tax collecting from the peasantry on behalf of the nobility. According to one document, in 1270, 80 percent of 228 adult Jewish males in Perpignan, France, made their living lending money to their gentile neighbors—a profession that put a premium on smarts or at least on a certain kind of intelligence.

History also suggests that a kind of sexual competition may have existed for centuries between wealthy and poorer Jews. Young Jewish scholars who were skilled in the verbal interpretation of the holy books would endlessly parse the minutiae of Jewish law and were by tradition selected for marriage to the daughters of wealthy Jews. Religious sanctions enforced this. The wealthiest Jews, presumably the smartest, would have the most children. Poorer Jews had a more difficult time attracting wives, so they usually married later and had fewer children. In other words, "the payoff to intelligence was indirect rather than direct." In evolutionary terms, fidelity and literacy paid off handsomely, creating a positive eugenic effect—the breeding of "smart" Ashkenazi Jews. Cochran also suspects that one of the important factors in increasing Ashkenazi IQ was simply the mixing of Jews with non-Jewish Europeans: hybrid vigor, geneticists call it. About 40 percent of the genes of Ashkenazim are European. Obviously you would need selective forces on top of that admixture to raise Jewish IQ even higher, Cochran added.

The unique historical experience of eastern and central European Jews might also explain why Ashkenazi Jews stand atop the IQ totem pole, while those Jews who lived for centuries in the Muslim world in the Mediterranean, North Africa, and Middle East—mostly Oriental Jews and some Sephardim—score so poorly on IQ tests, well below the world average of 100. In his controversial book *Race Differences in Intelligence*, Richard Lynn estimates the average IQ of Israel at about 95, held down in part by the non-Ashkenazi Jewish IQ mean of 91 and the 20 percent of the population that is Arab, with a mean IQ of 86 (equal to the median IQ of Arab countries surrounding Israel). The IQ of Ashkenazi Israelis tested at 103. (Critics of IQ tests claim that the low scores of non-Ashkenazim demonstrate the failure of these tests to accurately measure intelligence and the "cultural hegemony" of white Europeans, including many Ashkenazim, who design and administer them. It's proof, they say, that claims of high Jewish IQ are overblown.)

Cochran believes his theory can account for this IQ difference between Ashkenazim and other Jews. According to historians, there were far fewer taboos against Muslims lending money or collecting taxes. While a sizable minority of Jews in Europe thrived as rabbis, merchants, and moneylenders and had many children, the Jews of the Muslim world had a rougher go of it, barely getting by in "dirty jobs," such as cesspool cleaners, tanners, and hangmen. "The Jews of Islam, although reproductively isolated, seem not to have had the necessary concentration of occupations with high IQ elasticity," Cochran noted. Lynn believes the IQ differences could have been enhanced by selective survival (European Jews were more persecuted), selective migration (more intelligent Jews immigrated to Britain and the United States), and intermarriage with populations with different average IQs (Orientals married Middle Easterners and North Africans, while Ashkenazim married Europeans).

For this theory to hold true, it must be universal. Are there other isolated populations with high intelligence? This possible evolutionary trade-off shows up in at least one other ethnic group: the Parsis, a tiny Zoroastrian sect of less than eighty thousand people, most of whom live in India. The Zoroastrian religion dates to the Persian prophet Zarathustra, who preached around 1500 BCE. When Arab Muslims took control of Iran in the seventh century CE, a group of Zoroastrians fled to Bombay, where they became known as Parsis, derived from "Persians." The high priests of the Parsis had for centuries held that intermarriages would be suicidal for the minority community, which is less than 1 percent the size of the world Jewish community. They neither proselytized for their faith nor accepted converts.

Because of inbreeding, Parsis are plagued with Ashkenazi-like diseases—BRCA1 breast cancer, Parkinson's, tremor disorders, and mental retardation. In the Middle Ages, Parsis were outsiders in Hindu India, making their way as peddlers, brokers, and moneylenders and, after the Europeans arrived, as shipbuilders and traders. "By the mid-nineteenth century, the Parsis had become Bombay's leading bankers, industrialists, and professionals," wrote Yuri Slezkine, who draws parallels with the Jewish experience in *The Jewish Century*. There is anecdotal evidence that Parsis, like Ashkenazim, have maintained that high rate of artistic and business success—famous Parsis include the conductor Zubin Mehta and the Tata industrial family. The name of India's Gandhi dynasty traces not to Mahatma Gandhi but to Indira Nehru's Parsi husband. Gandhi once said of them: "In numbers, Parsis are beneath contempt, but in contribution, beyond compare." Sound familiar?

Cochran, Harpending, and Hardy seem to relish the controversy their theory has provoked, freely acknowledging that it "is closer to advocacy than it is to hypothesis testing." But they don't appear to have any ideological axes to grind. And none of them is Jewish. As they note, their theory is also testable: compare the IQs of sibling pairs, one of whom carries a brain disease gene and one of whom does not. If the authors are correct, disease sufferers with a single copy of a defective gene should be more intelligent than average—as many as five IQ points, they guess. If their thesis should hold up, in whole or in large part, it could revolutionize the way we interpret Western history.

But as my mother would have said, Is it good for the Jews? These data are available in Israel, which has huge banks of centralized medical records, but in view of the prickliness of the issue, Jewish researchers are not rushing to crack the numbers. Considering the Jewish ambivalence about stereotypes of genetic and cultural distinctiveness, it should also come as no surprise that Jewish intellectuals have been leading the attack on claims of innate Jewish intelligence.

Chief among them is the Emory University historian Sander Gilman. In his polemic on the debate over Jewish IQ, *Smart Jews*, Gilman denounces the "construction of Jewish superior intelligence" as the historical legacy of the projection of "intellectual queerness" by non-Jews onto Jews. "I'd actually call the study bullshit if I didn't feel its idea were so insulting," Gilman has said of the Cochran theory. He considers calling Jews innately smart a backhanded compliment that plays into stereotypes of Jews as devious, crafty, or athletic weaklings. Theories of groupwide genetic uniqueness understandably make people nervous. "I see no positive impact from this," said Neil Risch. "It's bad science—not because it's provocative but because it's bad genetics and bad epidemiology," agreed Harry Ostrer. When geneticists wade into the social quagmire of race, it rarely leads to any good.

But in my private conversations with dozens of geneticists, almost no one was willing to categorically rule out the theory that Ashkenazi Jews may have received a genetic gift of intelligence as recompense for the extraordinary number of brain diseases they are prone to. "Yes, selection for intelligence is credible," Hebrew University's Joel Zlotogora, one of the few scientists willing to be quoted, told me. "For me, everything is credible. I think the founder effect is true for only some disorders, but not for all of them, and there must be something else. There could very well be positive selection."

Duke University's David Goldstein had a similarly nuanced reaction. "Until recently, most human geneticists almost . . . disallowed discussion about genetic differences among racial and ethnic groups," he said. "So many awful things had been done with genetic research in this last century that they developed a policy of 'Just say no.' But there's actually a lot of difference between groups when you consider there are 10 million polymorphic sites on the genome. So it's not scientifically sound to rule out the possibility of differences corresponding to our geographic and ethnic heritages. It overlooks the basic point: The genome is just a huge place."

Remarkably few scientists were willing to talk on the record, especially in the United States, and no wonder. The groundbreaking study on intelligence genes by Bruce Lahn, the young geneticist at the University of Chicago, fired charges that some population groups may have a genetically based intelligence advantage. The outcry prompted him to abandon any new research. "It's getting too controversial," said Lahn. He believes that "society will have to grapple with some very difficult facts" as scientific data accumulates, but in the face of a firestorm, in his late thirties, he's thinking more about academic survival.

"It would be hard to overstate how politically incorrect [the Ashkenazi intelligence] paper is," said Steven Pinker, the Harvard University cognitive scientist, noting that it reopens the taboo discussion of race, genes, and intelligence. "It's certainly a thorough and well-argued paper, not one that can easily be dismissed outright."

"Postmodernists and specifically post-Zionists tend to emphasize the concept of constructed identities," noted Sergio DellaPergola, Israel's foremost demographer. "The theoretical possibility that given traits of human personality will be shown to have genetic bases, and that different groups might hold different probabilities of displaying certain (useful or harmful) characteristics, seems to these observers tantamount to opening Pandora's box."

Is it a risk worth taking? I asked him.

"Absolutely. Without question. Asking and answering such questions is what scientific inquiry is all about."

SCIENTIFIC RACISTS

We are in treacherous waters. This resurrection of racial biology for tracking ancestry, isolating diseases, and discovering ethnic traits has already begun to stir the pot of racism.

"I'm a scientific racist," Kevin MacDonald brashly told me. "Jews do not act in the best interest of society. We need to systematically put in place some controls, call it discrimination if you will, to restore parity with other groups."

MacDonald, who is in his sixties, is a psychologist at California State University at Long Beach, a nondescript-looking university with a reputation to match. I met up with him one sultry August day at his second-floor cubicle in a bland office park. It was tiny, but seemed even more constricted because of the magazines stacked staggeringly high in every nook and corner. He cleared a messy pile of books from the sofa chair, which was clearly not reserved for the rare visitor.

Excepting his ruddy face, he is a dead ringer for Ichabod Crane: tall and skinny, with slender, knobby hands, white hair, and a gawky demeanor. MacDonald is a proponent of sociobiology. According to classic natural selection theory, we are all programmed to produce and nurture our children, which ensures that our genes will be passed along to future generations. MacDonald and a small but growing group of evolutionary psychologists believe that humans will sometimes sacrifice their own self-interest for the greater good of the group. Some proponents of this theory believe this altruistic behavior is reflected in the gene pool of ethnic and racial groups.

MacDonald has written three explosive books analyzing the behavior of Jews and proposing a radical theory of Jewish separateness. He claims that Jews have an "evolutionary group strategy" that promotes Jewish interests at the expense of the greater (gentile) well-being. He traces Jewish separatism to biblical times but finds evidence of its greatest expression in the medieval ghettoes of eastern and central Europe. Using his jargon, Jews established an "extended kinship network" with "high levels of within-group cooperation and altruism." They were notable for "high investment parenting and commitment to group, rather than individual, goals." MacDonald concluded his first book on the Jews with a warning: "there is an urgent need to develop a scientific theory of Judaism and anti-Semitism."

Why? I asked him.

"Because right now Jews are very important. They're very important on the international scene because of Israel. I think we have to have a theory of Jews, and be able to discuss Jewish interests. I almost never see Jewish interests tackled; even Israel is never discussed as a Jewish issue. If it were more widely debated, perhaps people would be more honest about Jewish issues. I do see a difficult road ahead. I think there could be increasing ethnic conflict and polarization in this country."

MacDonald's account of Jewish "exceptionality" and the hubris that can accompany it is often persuasive, but his thesis reads like a genetically updated version of *The Protocols of Zion*: Jews have an almost diabolical, biologically programmed plan of dominance. Their group goal, imprinted in the genome over thousands of years, is to breed Jews with clever verbal skills that match well with another inbred Jewish trait, aggressiveness. MacDonald even finds a way to portray Jewish cultural assimilation and secularism as secret weapons in their group strategy. The Jewish promotion of multiculturalism is a charade. Even the anthropologist Franz Boas, who initiated the historic shift in anthropology from biology to culture, was supposedly motivated by a desire to end the criticism that Jews were a race so they could more easily pursue their group strategy of Jewish racial dominance. That's just the devious nature of Jews, he said.

MacDonald is of Scottish and German ancestry, although he calls himself German. He traces the evolution of his scientific racism to his college days. He says he was a political radical at the University of Wisconsin in the mid-1960s, when it was a hotbed of antiwar activism. "I had couple of Jewish roommates, and I got into the campus radical scene through them." He ended up becoming a hippie, taking drugs, and dropping out of school, moving to Boston, then to Berkeley, and later to Jamaica, where he taught high school, killing time playing jazz.

"I reflect a lot on that time," he said. "If I hadn't had those Jewish roommates, I never would have become involved in the radical movement. There were a lot of Jews whose parents had been radicals in the thirties. It struck me as odd because nobody from where I was from had those kinds of parents. For those Jews, they were living up to their parents' ideals. That was their culture."

MacDonald ultimately returned to finish his undergraduate degree and then went on to get a doctorate in developmental psychology at the University of Illinois. His academic interests took a turn in the mid-1970s, with the furor kicked up by the publication of Edward O. Wilson's *Sociobiology*. In the 1990s, with Wilson in mind, MacDonald turned his attention to Jews. "I think Jews are far more interesting than any other ethnic group," he said. "You've got the blacks, and it's sort of all predictable and pretty simple in a way. But with Jews, it's all the history, the rationalization, the point/counterpoint, and Jews thinking about themselves and ratio-nalizing themselves to non-Jews. Now you've got Israel. You have a situation over there that's viewed as a moral outrage by a lot of people, not just Palestinians. And to be able to pull it off and take the moral high ground in this country so that the average person on the street still thinks of Jews as really moral, as just really good people. I mean, how do you pull that off? Other groups haven't been able to do that."

Apart from an invidious air of Jewish conspiracy, MacDonald's writings present a selective slog through Jewish history, but with a twist. "Judeo-ethnocentrism" is the root cause of anti-Semitism and a general threat to society, he maintains. The Jews who were killed in ancient Rome, the Crusades, the pogroms, etc. were invariably zealots who all but invited their fate. Although he certainly doesn't praise the perpetrators, his books are suffused with the sense that Jewish intransigence made them complicit in their liquidation. Mac-Donald conjures the ghosts of Nazis past by modestly calling his obsession an "effort to develop a *Wissenschaft des Judentums*—a scientific understanding of Judaism," echoing the racist spark that produced the firestorm of National Socialism. Like Hitler, he be-lieves Jews are a racial threat to gentile society. The Holocaust was horrific but understandable—a counterreaction by gentiles to the attempt by Jews to control more than their share of resources; it just got out of hand.

MacDonald proposes a racial solution to bring the Jews to heel: reintroduce discrimination at prestigious universities and reinsti-tute quotas in the very high paying occupations they now excel in. He even proposes instituting special taxes on wealthy Jews to contain their power and influence.

He is not without credentials and some supporters. He is well known in evolutionary psychology circles and even edits an academic publication in the field. There is little written criticism of his work, however, in part because it's apparently largely gone unread, even by fellow evolutionary psychologists. Although he cogently reviews the history of Jewish victimology, his group theories and selective anecdotes suffer from an evidence problem. He systematically downplays Jewish scientific success, riffing on the Wagner-Hitler thesis of Jewish cleverness-disguised-as-accomplishment. He never persuasively addresses how the descendants of Spanish Jews could fall from the top of the heap to the basement in intelligence and achievement in less than sixteen generations and have that new standard of low achievement imprinted in their non-Ashkenazi genes. And like some scholars who invoke evolutionary psychology, he presents almost no quantitative or experimental data to support his speculation.

When the genetic research on Jewish distinctiveness began coming out in the late 1990s, MacDonald could not have been more pleased. "It fits very well with my theories," he claimed. "It suggests that Jews have something genetically unique," which is of course accurate, but not necessarily in the way he has portrayed it in his writings. He has always been struck by the fact that Jews have welcomed their self-image as separatists, which he thinks is now programmed into their DNA. "It is the very fabric of Jewishness." Although Judaism actively embraced converts in its early history, that door almost closed over the course of Jewish history. "If the rabbis could have done so, they would have made the walls even higher," he said, quoting Paul Johnson, the respected author of *The History of the Jews*. "The issue of conversion and intermarriage is a hot-button issue. When I reviewed that, I concluded that the level of intermarriage and conversion was really small. The ones who were admitted tended to be nonreproductive. They were going to the very lowest levels of Jewish society." By his guess, the few converts did not "pollute" the "Jewish gene pool."

As even he admits, his dark "science of Jewry" rests precariously

H	Linked to population expansion about 20,000 years ago. Originated in the Caucasus or in Europe and now found in 50 percent of those of European ancestry, and common in North Africa and the Middle East.
I	Believed to have arisen in Eurasia some 30,000 years ago and one of the first haplogroups to move into Europe. It's the lineage of the 5,000-year-old Ice Man and is found mostly in southern Europe, but also in Egypt and Arabia.
J, J1, J2	Originated about 45,000 years ago in central Asia and is associated with the spread of farming and herding in Europe during the Neolithic Period, beginning 10,000 years ago. Common in the Near East, Europe, the Caucasus, North Africa, and the Middle East and among Jews. J2 is more localized in the Mediterranean.
K	Subbranch of U that is believed to have first appeared in the early stages of the Holocene Epoch, when populations expanded into Europe after the last glacial maximum. About one-third of people with Ashkenazi ancestry carry a subclade of this haplogroup.
L1, L2, L3	Eve's ancestors in southern Africa whose first daughters settled in separate communities in West Africa and northwestern Africa. L1 is the earliest mitochondrial lineage, appearing 145,000 to 170,000 years ago. Common in sub-Saharan Africa, especially in the Khoisan people, L2 is the main ancestral link to all lineages outside of Africa. L3 is confined mostly to East Africa and emigrant African populations.
M	With N, one of two "superhaplogroups." Offshoot of L3, the descendants of this founding mother followed the Rift Valley into the Middle East, then spread to India, southern and eastern Asia, and on to Australia. More than half of Indians have this haplogroup. Believed to have spawned haplogroups C, D, E, G, and Z.
N	With M, one of two "superhaplogroups." It's the daughter of one of the northeastern African lineages and is believed to have originated in Africa 60,000 years ago. Common in the Middle East and a primary source of the various European female lineages, including B, F, H, J, R, T, and U.

N1b	Present, although in low frequencies, in Europe, the Caucasus, the Near East, Egypt, and Arabia. Radiated around 39,000 to 52,000 years ago, spawning at least four ancestral clusters, including haplogroup B, that expanded to eastern Asia, reaching Japan and the southeastern Pacific archipelagos.
T	Believed to have originated in Mesopotamia or Anatolia 10,000 years ago and to have moved northward with the spread of farming. Found most commonly around the eastern Baltic Sea and in the Urals.
U	Very large and diverse branch. Dates to approximately 50,000 years ago, when it may have interacted with Neanderthals living in Europe. Found in Basques, the Welsh, and a handful of Atlantic coastal populations and was the mtDNA of Cheddar Man. Also links back to Africa. About 40 percent of Europeans have the clade U5, the most frequent and ancient subcluster of U, while U2 is most common in India and U6 in North Africa.
V	Thought to have arisen about 12,000 years ago, possibly in Iberia. Found in high concentrations in the Saami population of northern Scandinavia, as well as in the Basque people.
W	Appears in the western Ural Mountains and the eastern Baltic area, though it is also found in India.
X	An offshoot of M and N, this lineage traveled through the northern Levant into Europe. Intriguingly, X is also found in northern Asia and is believed to have arrived in the Americas about 15,000 years ago, which would predate the primary Native American lineages and make this line present among the first settlers in the Americas. A number of Indian groups, including the Sioux, have the X haplogroup. There is speculation that the Caucasian-like skeleton known as Kennewick Man might have this lineage.
Z	Descended from M, believed to have arisen in central Asia and now found in Korea, northern China, central Asia, and Russia.

ADAM/Y-CHROMOSOME HAPLOGROUPS

Like the mtDNA map of haplogroups, Y haplogroups have been assigned alphabetical designations.

Y LINEAGES

A	Believed to have been the first Y-chromosome lineage to diverge. Restricted to Africa, where it is present in several populations at low frequency, but is most commonly found in populations of the Khoi and the San tribes of southern Africa.
B	One of the oldest Y-chromosome lineages in humans. Found exclusively in Africa, most commonly in Pygmy populations. First lineage to disperse around Africa. There is current archaeological evidence supporting a major population expansion in Africa approximately 90,000 to 130,000 years ago, which may have spread this lineage throughout Africa.
C, C3	Found throughout mainland Asia, the South Pacific, and at low frequency in Native American populations. Originated in southern Asia and spread in all directions. This lineage colonized New Guinea, Australia, and northern Asia, and currently is found with its highest diversity in populations of India.
D	Originated in Africa 50,000 years ago and shares a common ancestry with haplogroup E. Represents a great coastal migration along southern Asia, from Arabia to Southeast Asia, and northward to populate East Asia. It is found today at high frequency among populations in Tibet, the Japanese archipelago, and the Andaman Islands, though not in India.
D2	Most likely derived from the D lineage in Japan. Completely restricted to Japan and is a very diverse lineage within the aboriginal Japanese and in the Japanese population around Okinawa.
E	Originated in Africa 50,000 years ago and shares a common ancestry with haplogroup D. Contains a number of groups in Africa and the Middle East. Some derived clades are also found at moderate frequencies in the populations of Europe, particularly among those that reside near the Mediterranean, which is believed to represent ancient genetic influence from the Middle East to Europe.

E3a	African lineage that most likely dispersed south from northern Africa within the last 3,000 years with the Bantu agricultural expansion. The most common lineage among African Americans.
E3b	Believed to have evolved in the Middle East before spreading into the Mediterranean during the Pleistocene Neolithic expansion. Found in many Arab populations and in areas around the Mediterranean; in East and North Africa, particularly among the Berbers; and in southeastern Europe.
F	First appeared in Africa some 45,000 years ago. It is believed to represent the second wave of expansion out of Africa. F* is the ancestral haplogroup of the Y-chromosome haplogroups.
G, G2	May have originated along the eastern edge of the Middle East or in India or Pakistan 30,000 years ago and has dispersed into central Asia, Europe, and the Middle East. The G2 branch of this lineage (containing the P15 mutation) is found most often in the Caucasus, the Balkans, Italy, and the Middle East.
H	Arose 20,000 to 30,000 years ago and is almost entirely restricted to India, Sri Lanka, and Pakistan, where it probably originated.
I	The I and its various subclade lineages are concentrated in Scandinavia and Croatia, with some traces in the Middle East, its probable source. These would most likely have been common within Viking populations.
J, J2, J1	Arose 10,000 to 15,000 years ago in the Fertile Crescent. Includes Jews, Arabs, Armenians, and Kurds. Found at its highest frequencies in Iran and Iraq, from where it most likely originated, and then was carried by traders into Europe, central Asia, India, and Pakistan. J2, which is believed to be associated with the spread of agriculture during the Neolithic Period from Anatolia, is found throughout central Asia, the Mediterranean, and south into India. While the majority of haplogroup J is not Jewish, the majority of Jewish men fall into J. The Cohen Modal Haplotype is found in haplogroup J1.

K, K*	First appeared 40,000 years ago in Iran or southern central Asia and was the haplogroup of the patrilineal ancestors of most of the people living in the Northern Hemisphere, including most Europeans, many Indians, and almost all Asians. Other lines derived from K* are found among Melanesian populations, indicating an ancient link between most Eurasians and some populations of Oceania.
L	Common in India, where it likely originated 30,000 years ago.
M	It is believed to have first appeared 10,000 years ago and is found mostly in Southeast Asia, particularly Melanesia, Indonesia, and Micronesia.
N	Distributed throughout northern Eurasia, most commonly in Finns, Russians, and Hungarians. Likely originated in northern China or Mongolia and then spread into Siberia, where it became a very common line in western Siberia.
O, O3	Believed to have originated in Siberia 35,000 years ago. Appears in 80 to 90 percent of all human males in East and Southeast Asia and is almost exclusive to that region. O3 shows up in more than 50 percent of Chinese males and about 40 percent of northeastern Asians in Manchuria and Korea.
O1	Found at very high frequency in the aboriginal Taiwanese (possibly due to genetic drift). Probably originated in East Asia and later migrated into the South Pacific. Individuals carrying this lineage are thought to have been important in the expansion of the Austronesian language group into Taiwan, Indonesia, Melanesia, Micronesia, and Polynesia.
P	Believed to have originated in Eurasia 35,000 to 45,000 years ago, it's found throughout Europe and in almost all the indigenous peoples of the Americas.
Q	Links Asia and the Americas. It is found in northern and central Asian populations as well as in Native Americans. Believed to have originated in central Asia 15,000 to 20,000 years ago and migrated through northern Eurasia into the Americas.

Q3	Strictly associated with Native American populations. Defined by the presence of the M3 mutation (also known as SY103), which occurred 8,000 to 12,000 years ago, as the migration into the Americas was under way. There is some debate about which side of the Bering Strait this mutation first occurred on, but it definitely happened in the ancestors of the Native American peoples.
R, R1, R2	Believed to have originated somewhere in northwestern Asia between 30,000 and 35,000 years ago. R1 is very common throughout Europe and western Eurasia. R2, which is much rarer, is found only in Indian, Iranian, and central Asian populations.
R1a	Most common among eastern European Slavs and in populations in India and central and western Asia. May be the Khazarian Jewish lineage. Thought to have originated in the Eurasian steppes north of the Black and Caspian seas in the Kurgan people, believed to be the first speakers of the Indo-European language group, who lived in the region around 3000 BCE and are known for the domestication of the horse.
R1b	Most common haplogroup in European populations. Believed to have expanded throughout Europe as humans recolonized after the last ice age, 10,000 to 12,000 years ago. Contains the Atlantic Modal Haplotype.

APPENDIX 3

TRACING YOUR ANCESTRY AND FAMILY HISTORY USING DNA

Our personal histories—what genealogists call our "deep ancestry"—are often beyond the discovery capabilities of most genealogical research. Many populations or ethnic groups, such as African Americans and Jews, who have lived through genocides and spent centuries in diasporas, have found their genealogical heritage blurred or erased. Others of us have just lost touch with our family stories or are curious about our ancient roots. Now population genetics provides an increasingly sophisticated tool for probing ancestral vaults once thought lost to history.

Genetic genealogy is fast becoming a booming industry. You are now able to explore your European, Native American, Asian, African, or Jewish ancestry, among other lineages. Home DNA test kits can be ordered through the mail or over the Internet at a cost ranging from less than $100 to $1,000 or more for a variety of ancestral tests. They come with a swab to easily collect a sample of cells from the inside of your mouth. You send back the sample through the mail, and within a short time, you receive the results—a series of numbers that represent key chemical markers on your DNA and an explanation of what those numbers reveal about your ancestry.

Molecular genealogy uses genetic markers to link people together into family trees, tribal groups, ethnic groups, and what are popularly known as races. Although we are almost identical at the level of our DNA, it is those minuscule differences that impart our individuality, signal how genetic ethnic groups have formed and dispersed, and provide evidence of our distant ancestry. By reading the mutations that differ between the world's various populations, it is possible to make a strong inference about our ancestral mix and identify common ancestors. The mutations that appear on the autosomes, the sex chromosomes, and mtDNA provide the data for geneticists to calculate your ancestral lineage or clan on your paternal and maternal sides and which modern population you most resemble genetically. They read a section of your DNA and compare its sequences to thousands of others from all over the world.

DNA is the carrier of our genetic information, which passes from generation to generation. At conception, a person receives DNA from both his or her father and mother. We each have twenty-three pairs of chromosomes, and for each pair, one was contributed by the father and another by the mother. These twenty-three pairs of chromosomes are known as nuclear DNA, since they reside in the nucleus of every cell, except red blood cells. Twenty-two of the chromosomes are known as autosomes. One half of the twenty-third chromosome, from the mother, is always an X. From the father, a person either inherits an X chromosome or a Y chromosome, which determines the sex of the child. Getting an X from the father would result in an XX, who would be female, and getting a Y from the father would result in an XY, who would be male. Both males and females inherit mitochondrial DNA, located in abundant quantities outside the nucleus of each cell, from the mother. The father does not pass on any mitochondrial DNA.

DNA tests are fascinating, particularly to those interested in genealogy, but they have limitations in what they can tell us. That's because the tests show only small slices of our genetic history. The two most popular tests focus on our male and female lineages, but each of us has many, many more ancestral lines in our jumbled histories. If you go back just ten generations, or 250 to 300 years, you will have 1,024 ancestors.

There are three basic DNA tests now used by genealogists: Y-chromosomal DNA tests, which trace the paternal line; mitochondrial DNA (mtDNA) tests, which document the maternal pedigree; and so-called human identity tests, which use autosomal DNA, markers on the nonsex chromosomes, which can help tell us our "racial mix." Anyone can take an identity test. MtDNA tests that determine maternal ancestry can be done on both males and females, because both sexes inherit mtDNA from the mother. Because the Y chromosome is found only in men, those who take the Y DNA test must be males. For females who are interested in the Y DNA result for their surname or family tree, they would request that a close male relative participate, such as a brother, father, uncle, or cousin with the same surname as the father, to ensure that the line of descent is correct. DNA tests can help ascertain

- whether you and another person are related either recently or in the distant past (you share a common ancestor) (Y DNA and mtDNA tests);
- whether a set of men with the same or similar surname are directly related through a common ancestor (Y DNA);
- the genetic origins or country of the ancestors of your population group (Y DNA for male lineage, mtDNA for female lineage).

Three major services—Family Tree DNA, Oxford Ancestors, and Relative Genetics—offer tests of both male and female lineages. Overall, the best package of services is provided by Family Tree DNA, which pioneered the use of Y-chromosomal testing in genealogy and uses the most markers, which significantly enhances the reliability of the results. FTDNA is affiliated with Michael Hammer's world-renowned genetic anthropology laboratory at the University of Arizona, which identified the Cohen Modal Haplotype and is where all its samples are analyzed. It tests about twenty thousand people a year. Anyone interested in finding Jewish or Semitic roots would find the database at FTDNA the most extensive by far for finding common ancestors. Here is a list of some of the major genetic genealogy services:

FAMILY TREE DNA
1919 North Loop West, Ste. 110
Houston, TX 77008
Phone: (713) 868-1348
Fax: (713) 868-4584
http://www.familytreedna.com
info@FamilyTreeDNA.com

A collaboration of Bennett Greenspan and the Michael Hammer team at the University of Arizona, Family Tree DNA is a testing service that also assists with genealogical research, including surname searches, and offers a time predictor for the likely date of the most recent common ancestor of two related individuals. It also offers family migration and Native American territory maps.

AFRICAN ANCESTRY
5505 Connecticut Ave., NW
Box 297
Washington, DC 20015
http://www.africanancestry.com
info@africanancestry.com
Phone: (202) 723-0900
Fax: (202) 318-0742

African Ancestry offers matrilineal and patrilineal tests for descent from specific countries of Africa and/or ethnic groups using a database weighted toward the parts of the continent where slave traders operated.

OXFORD ANCESTORS
PO Box 288
Kidlington, Oxfordshire, UK
OX5 1WG
Phone: 44 01865 374 425
http://www.oxfordancestors.com
enquiries@oxfordancestors.com

Founded by Bryan Sykes, Oxford Ancestors offers Y-chromosome and mtDNA searches, with a specialty in the tribes of Britain and the matrilineal and patrilineal lines of Europe.

RELATIVE GENETICS
2495 South West Temple
Salt Lake City, UT 84115
Phone: (801) 461-9760
Fax: (801) 461-9761
http://www.relativegenetics.com
info@relativegenetics.com

Relative Genetics offers testing of paternal and maternal lineages and tracks ancestral origins using autosomal genetic markers.

TRACE GENETICS
4655 Meade St., Ste. 300
Richmond, CA 94804
Phone: (866) 731-2312
Fax: (510) 233-5336
http://www.tracegenetics.com
info@tracegenetics.com

Trace Genetics is a professional forensic and anthropologic laboratory that offers genetic genealogical services, with a specialty in Native American DNA.

Y DNA AND MTDNA TESTS

All Y and mtDNA tests allow you to identify your ethnic and geographic origins, both recent and far distant, on your direct male or female line of descent. Scientists have identified small portions of the Y and mtDNA that are passed on virtually unchanged from father to son or mother to daughter. Testing of this portion provides information about direct lineages by identifying genetic mutations—at markers or loci—that occur every few hundred generations per marker. At a locus, there might be ten to thirty alleles, or identification markers, that are valuable for genetic anthropologists or genealogy study. The markers on the Y chromosome, known as short tandem repeats (STRs), have been assigned numbers (e.g., DYS#391) set by the Human Gene Nomenclature Committee, which is part of the Human Genome Organisation (HUGO), based at University College London. Not all loci used in DNA reports have yet been assigned universal numbers. Some designations are proprietary and assigned and used by individual commercial forensic or genealogical laboratories.

Generally, the greater number of markers looked at, the more precise the DNA test and the more likely that a most recent common ancestor (MRCA) can be identified in recent generations. The results will not be able to indicate who this specific ancestor is but may be able to help you narrow it down to within a few generations, which can be important for genealogical research. Population geneticists then apply a term known as the most likely estimate (MLE) of how many generations ago your MRCA would have lived. DNA tests can provide only a rough estimate, and in each individual case, the actual generation could be nearer or further from the person tested. In their calculations, geneticists use twenty-five years to represent a typical generation before the Middle Ages and twenty-five to thirty years per generation thereafter. Because scientists do not yet know the mutation rates for specific mutations, testers currently use the mutation rate averages suggested by geneticists for each mutation event: for the Y chromosome—one mutation per five hundred generations, or 10,000 to 12,500 years per locus (marker); for mtDNA—one mutation every three hundred to six hundred generations, or 6,000 to 15,000 years per marker. Each locus may have a different rate of mutation, some faster and some slower.

On the male chromosome, if you match another person exactly in a dozen markers—a rare occurrence, unless you are closely related—you have a 99 percent likelihood of sharing common ancestry. Many people who share a surname will share a perfect or near-perfect match. Many surnames are much older than a few hundred years, and two people may share a surname but not one or two of their markers. In these cases, the MLE of when their MRCA lived could be further back in time. Both the Y-chromosomal and mtDNA tests create a profile, known as a haplotype, that distinguishes one male-to-male or female-to-female lineage from another. If your alleles differ in one locus from another person by just one position (one mutation event, also known as a step or a point), you would be considered in the same cluster of interrelated haplotypes.

Using more markers improves your ability to determine how recently and therefore how closely you are related. For example, if two individuals match exactly at all loci in a twelve-marker test, there is a 50 percent probability of an MRCA within the last seven generations. If they match exactly at all loci in a twenty-five-marker test, there is a 50 percent probability of an MRCA within the last five generations. With all thirty-seven markers matching, there is a 50 percent probability that the MRCA was no more distant than three generations ago and a 90 percent probability that the MRCA was within the last sixteen generations. Of the services, only Family Tree DNA offers a thirty-seven-marker Y-chromosomal test.

Haplotypes that cluster together in a definable population are known as a haplogroup, which can provide information about the ancient origin of a lineage. Your haplogroup defines where the first male or female in your lineage originated. Many haplogroups are continent- or ethnic-specific, and subdivisions of these haplogroups may be region-specific. Each mtDNA haplogroup originated in a different geographic area thousands or tens of thousands of years ago. The classification of mtDNA haplogroups should not be confused with the classification or region of origin of Y DNA haplogroups, even though they use identical alphabetical designations (A, B, C, etc.). Y DNA haplogroups are based on the Y chromosome, and mtDNA haplogroups are based on mitochondrial DNA, which makes them unrelated in any way. As compared to the Y-chromosome tree, mtDNA is much more geographically scattered. Consequently, the human mitochondrial map is more complex than the Y DNA tree.

A NOTE ABOUT HUMAN IDENTITY TESTS

These tests, which purport to measure your genetic ethnicity, or "race," by testing a number of markers spread throughout your twenty-two autosomal pairs, have been widely used in forensic casework since 1993. Using complex statistical algorithms, the tests match people to one of four major biogeographical ancestral groups—sub-Saharan Africans, Indo-Europeans, East Asians, or Native Americans—and approximate the relative percentages of admixture. For example, one person may obtain a result of 80 percent European and 20 percent Native American, while another may type as 33 percent African, 15 percent Native American, and 52 percent European.

Because the tests use only a tiny fraction of the potential autosomal sequences in the vast sea of the human genome, they are considered crude and sometimes misleading. Among other problems, the results do not readily distinguish between recent admixture and ancient mixing. So, someone with a significant percentage of, for example, African ancestry might have inherited those DNA fragments tens of thousands of years ago, while another person may have gotten his or her African DNA in recent generations. Greeks, Italians, Middle Easterners, and Jews systematically show low levels of Native American admixture, even with no known Native American ancestors. One company, Ancestry by DNA, offers this proprietary service. According to its Web site, http://www.ancestry-bydna.com, "customers receive a CD with their raw genetic data, a bar graph showing the percentages of each group, and a specialized representation of their data called a triangle plot, along with a user's manual."

Appendix 4

Case Study: The DNA of Father William Sánchez

William Sánchez's amazing quest to unravel his family's ancestry, detailed in chapter 2, took flight when he requested a test kit from Family Tree DNA. The findings were presented to him in a number of charts, which with Father Sánchez's permission are reproduced to provide some idea of the process.

Y DNA

The Y test yields a snapshot of "deep ancestry"—ten thousand or more years ago. Father Sánchez's Y tree appears in figure A4.1.

According to Bennett Greenspan, the founder and president of Family Tree DNA, Father Sánchez's paternal line is clearly Semitic as can be seen from his haplogroup tab, which shows matches with both Jews and Arabs. Haplogroup J is found at its greatest genetic diversity in the Zagros Mountains in western Iran. Populations of Iraq and Syria are substantially haplogroup J (55+ percent). More than 50 percent of Ashkenazi Jews are in haplogroup J (either J1 or J2), and similar or greater percentages are found in Jews from Yemen, who have been separated from the bulk of world Jewry for at least 2,500 years. The Buba clan within the

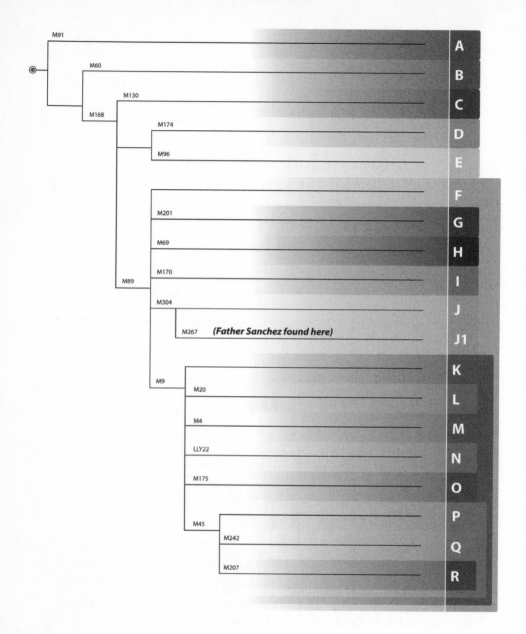

Figure A4.1. Father Sánchez's Y tree. (Courtesy of Family Tree DNA.)

African Lemba is an ancestral subset of the Yemenite Jews whose signature is also on J1.

The discovery that Father Sánchez's paternal line is haplogroup J1 is somewhat surprising, as Spain has no large population group from within that subhaplogroup—in fact, only about 7 percent of the Spanish population tests as J1. Given the relatively small percentage of this lineage in the rest of Europe, it can safely be assumed that many J1's from Spain share a crypto-Judaic past, which is probably true of Father Sánchez. It is also interesting to note that the signature of the Cohanim—descendants of Jewish priests—is also found in haplogroup J1. Family Tree DNA provides the actual scientific allele values and short tandem repeat designations for each of sixty-seven markers, which allows for more precise genealogical mapping, particularly as more genetic anthropology studies identify more source populations.

William Sánchez

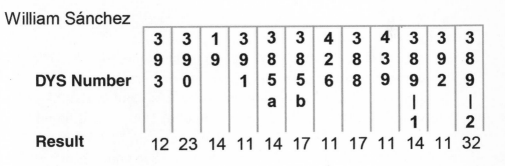

DYS Number	393	390	19	391	385a	385b	426	388	439	389 I 1	392	389 I 2
Result	12	23	14	11	14	17	11	17	11	14	11	32

Figure A4.2. Father Sánchez's allele chart. (Courtesy of Family Tree DNA.)

FTDNA also provides information about one's recent ethnic origins. Over generations, people tend to move, as do borders, so nationality or ethnicity is often blurred. That's true of diaspora populations, like Jews or the Scots who moved to Ireland in the 1600s, and of many other populations, such as the European immigrants who flooded into the United States at the beginning of the twentieth century. FTDNA compares the STR markers against its database (and Professor Hammer's) to find exact matches or nearly exact matches. A near match is either one or two steps away. If someone mismatches with another person by one or two steps (mutations), he or she is closely associated with that person genetically, but their genealogical connection is probably further back in time. The closer the match, the more recently the two people likely shared a common ancestor. An exact match is twelve out of twelve, twenty-five out of twenty-five, thirty-seven out of thirty-seven, or sixty-seven out of sixty-seven; a one-step match is eleven out of twelve, twenty-four out of twenty-five, or thirty-six out of thirty-seven, etc. Near matches show where those who are distantly related to a person have migrated over time. All of Father Sánchez's exact matches are from Spain and/or of likely crypto-Jewish ancestry.

MTDNA

Father Sánchez's maternal lineage is reproduced in his mtDNA family tree (figure A4.3).

Father Sánchez's mother is descended from haplogroup B. This lineage originated in eastern and southeastern Eurasia and is now found in that area and throughout the Americas—and in about 26 percent of FTDNA's Native American samples. This group is found no further north in North America than the Great Plains and no further south in South America than central Chile and Argentina. It is currently presumed that all Native Americans came to the New World from the Altai Mountains area of central Asia about 15,000 years ago. This particular haplogroup was present in the populations that initially colonized the pre-Columbian Americas and dates to at least 12,500 years ago. It's found in many Mexicans and southwestern Hispanos who have European or other non–Native American signatures on their fathers' side. Many crypto-Jews who relocated to the Americas were men who married local women after their arrival—a pattern common in Jewish migrations throughout history, according to mtDNA and Y-chromosomal studies.

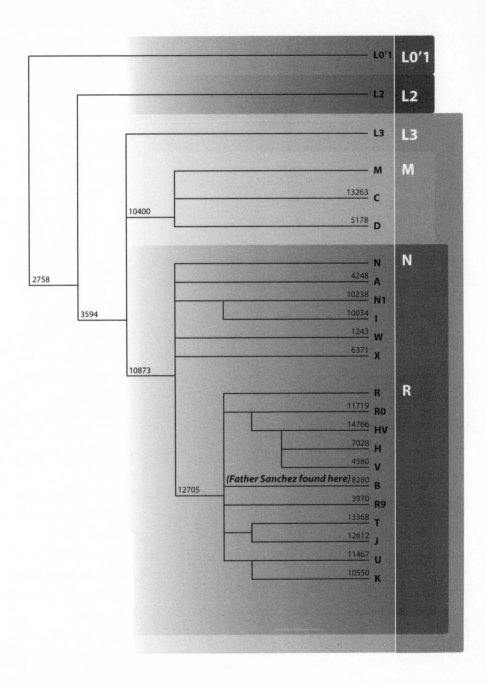

Figure A4.3. Father Sánchez's mtDNA tree. (Courtesy of Family Tree DNA.)

Appendix 5

"Jewish Diseases"

Ashkenazi diseases include:

- Abetalipoprotienemia (Bassen-Kornzweig syndrome)
- APC (adenomatous polyposis coli)
- Bloom syndrome
- Breast cancer and ovarian cancer (BRCA1 and BRCA2)
- Canavan disease
- Colorectal cancer due to hereditary nonpolyposis (HNPCC)
- Congenital adrenal hyperplasia (nonclassical form)
- Crohn's disease
- Cystic fibrosis
- Familial dysautonomia (Riley-Day syndrome)
- Familial hypercholesterolemia
- Familial hyperinsulinism
- Familial Mediterranean fever
- Fanconi anemia
- Gaucher disease type 1 (chronic adult noncerebral form)
- Glycogen storage disease, type 1
- Hemophilia C
- Lipoamide dehydrogenase deficiency
- Mucolipidosis IV
- Nonsyndromic hearing loss

- Niemann-Pick disease type A (acute neuropathic form)
- Parkinson's disease
- PTA deficiency (Plasma thromboplastin antecedent, or Factor XI, deficiency)
- Spongy degeneration of the central nervous system
- Tay-Sachs disease
- Torsion dystonia
- Ulcerative colitis
- Von Gierke disease
- Werner syndrome

Diseases common to Sephardim, Oriental Jews, and other Jewish populations include:

- Ataxia-telangiectasia (Moroccan Jews)
- Autoimmune polyglandular syndrome (Iranian Jews)
- Cerebrotendinous xanthomatosis (Moroccan Jews)
- Complement C7 deficiency (Moroccan Jews)
- Congenital adrenal hyperplasia (Sephardim, Moroccan Jews)
- Congenital myasthenia gravis (Iranian Jews)
- Corticosterone methyl oxidase type II deficiency (Iranian Jews)
- Creutzfeld-Jacob disease (Libyan Jews)
- Cystinosis (North African, Iraqi Jews)
- Cystinuria (Libyan Jews)
- Deubin-Johnson syndrome (Iranian Jews)
- 11-ß hydroxylase deficiency (Moroccan Jews)
- Familial hypercholesterolemia (Sephardim)
- Familial Mediterranean fever (Sephardim, Roman Jews)
- Glanzmann thrombasthenia (Iraqi Jews)
- Glucose-6-phosphate dehydrogenase deficiency (Kurdish Jews)
- Glycogen storage disease, type III (Moroccan Jews)
- Hereditary inclusion body myopathy (Iranian Jews)
- Limb-girdle muscular dystrophy with inflammation (Yemenite Jews)
- Metachromatic leukodystrophy (Habbanite Jews)

- Oculopharyngeal muscular dystrophy (Bukharan Jews)
- Phenylketonuria (Yemenite Jews)
- Tay-Sachs disease (Moroccan Jews)
- Tel Hashomer camptodactyly syndrome (Sephardim)
- Thalassemias (Kurdish Jews)

NOTES

Chapter 1. The Dead Sea Scrolls of DNA

P. 6 "risk for women . . .": King, Mary-Claire et al., "Breast and Ovarian Cancer Risks Due to Inherited Mutations in *BRCA1* and *BRCA2*," *Science* 302, October 24, 2003, 574–575. There have been more than twenty similar but less extensive studies, some of which have yielded lower estimates of the breast cancer risk to women, averaging 65 percent for BRCA1 and 45 percent for BRCA2.

P. 9 "no existing records . . .": Finkelstein, Israel and Silberman, Neil Asher, *The Bible Unearthed* (New York: Free Press, 2001), 48–71.

PP. 11–12 "thousands of genetic disorders . . .": Risch, Neil et al., "Categorization of Humans in Bio-Medical Research: Genes, Race and Disease," *Genome Biology* 3, 2002, 1–12.

Chapter 2. Identity Complex

P. 15 "'Christianity may have changed . . .'": William Sánchez interview, April 3, 2004.

P. 17 "'Believing in his own chosenness . . .'": Arendt, Hannah, *The Origins of Totalitarianism* (New York: Harvest Books, 1973), 73–74.

PP. 19–20 "Approximately 50 percent of Jews . . .": DellaPergola, Sergio, "The Jewish People 2004: Between Thinking and Decline," *Jewish People Policy Institute*, presented to the Israeli cabinet on July 4, 2004.

P. 26 "Many fled for refuge . . .": Kelly, David, "DNA Clears the Fog Over Latino Links to Judaism in New Mexico," *The Los Angeles Times*, December 5, 2004.

P. 26 "'We always thought . . .'": Norbert Sánchez interview, April 4, 2004.

P. 27 "'My mother never talked about it . . .'": Maria Sánchez interview, April 3, 2004.

Chapter 3. Blood Ties

P. 32 "Most scholars believe Ur . . .": "Against the Tide: An Interview with Maverick Scholar Cyrus Gordon," *Biblical Archaeology Review* 26, November/December 2000; Shanks, Herschel, "Abraham's Ur: Is the Pope Going to the Wrong Place?" *Biblical Archaeology Review* 26, January/February 2000.

P. 32 "Mesopotamia in the fifth . . .": Cooper, Jerrold S., "Sumerians" in *The Anchor Bible Dictionary* Vol. VI (New York: Doubleday, 1992), 232–233.

P. 39 "The Bedouin are camel breeders . . .": Knauf, Ernst Axel, "Bedouin and Bedouin States" in *The Anchor Bible Dictionary* Vol. I (New York: Doubleday, 1992), 637.

P. 43 "blood types of Jews . . .": Diamond, Jared, "Who Are the Jews?" *Natural History* 102, November 1993, 14.

PP. 43–44 *"The History and Geography of Human Genes . . .":* Cavalli-Sforza, Luigi L., Menozzi, Paolo, and Piazza, Alberto, *The History and Geography of Human Genes* (Princeton: Princeton University Press, 1994).

P. 45 "Cavalli-Sforza cites the case . . .": Cavalli-Sforza, Luigi L. and Cavalli-Sforza, Francesco, *The Great Human Diasporas* (Reading, MA: Addison-Wesley Publishing, 1995).

P. 49 "Sarich and Wilson concluded . . .": Sarich, Vincent and Wilson, Allan, "Immunological Time Scale for Hominoid Evolution," *Science* 158, 1967, 1200–1203.

Chapter 4. Eve and Adam

P. 53 "Cann and Stoneking . . .": Cann, Rebecca L., "Genetic Clues to Dispersal in Human Populations: Retracing the Past from the Present," *Science* 291, March 2, 2001, 1742+.

PP. 54–55 "'the Y chromosome . . .'": Jobling, Mark A. and Tyler-Smith, Chris, "The Human Y Chromosome: An Evolutionary Marker Comes of Age," *Nature Reviews Genetics* 4, August 2003, 598–612.

P. 55 "The breakthrough . . .": Mullis, K.B., "The Unusual Origin of the Polymerase Chain Reaction," *Scientific American* 262, 1990, 56–65.

P. 56 "Hammer realized . . .": Hammer, Michael, "A Recent Common Ancestry for Human Y Chromosomes," *Nature* 378, 1995, 376–378.

P. 57 "The first three sons of Adam . . .": Underhill, Peter A. et al., "Y Chromosome Sequence Variation and the History of Human Populations," *Nature Genetics* 26, November 2000, 358–361.

P. 57 "the case of the death of the last Russian czar . . .": Stevens, Richard F., "The

History of Haemophilia in the British Royal Family," *British Journal of Haemotology* 105, 1999, 25–32.

PP. 57–58 "One of the more amusing applications . . .": Sykes, Bryan, *The Seven Daughters of Eve* (New York: Norton, 2001).

PP. 58–59 "*Nature* published 'Jefferson Fathered Slave's Last Child,' . . ." Foster, Eugene A. et al., "Jefferson Fathered Slave's Last Child," *Nature* 396, November 4, 1998, 27–28.

P. 59 "Some still challenge . . .": http://www.angelfire.com/va/TJTruth/, maintained by a Jefferson family historian, Herbert Barger, is a clearinghouse for information and arguments against any conclusion linking Jefferson and Hemings as other than master and slave.

P. 59 "DNA has ruled out . . .": www.monticello.org/plantation/hemingscontro/hemings resource.html, maintained by The Thomas Jefferson Foundation, summarizes the studies and debate.

P. 59 "Y studies have also . . .": Zerjal, T. et al., "The Genetic Legacy of the Mongols," *American Journal of Human Genetics* 72, March 2003, 717–721.

PP. 59–60 "'The alternative explanation . . .'": Sailer, Steve, "Genes of History's Greatest Lover Found?" United Press International, February 5, 2003.

P. 60 "the Manchu warrior Giocangga . . .": Tyler-Smith, Chris, "Recent Spread of a Y-Chromosomal Lineage in Northern China and Mongolia," *American Journal of Human Genetics* 77, December 2005, 1112–1116.

P. 60 "A team of researchers at Trinity . . .": Moore, Laoise T., "A Y-Chromosome Signature of Hegemony in Gaelic Ireland," *American Journal of Human Genetics* 78, January 2006, 334–338.

P. 61 "Twenty-five generations ago . . .": Olson, Steve, "We're All Related to Kevin Bacon," *The Washington Post*, December 8, 2002, B2.

P. 61 "Many millions . . .": Rohde, Douglas L.T. et al., "Modelling the Recent Common Ancestry of All Living Humans," *Nature* 431, September 2004, 562–566.

P. 62 "'It's really dangerous to market . . .'": Shute, Nancy, "Where We Came From," *US News and World Report*, January 29, 2001.

P. 62 "'For me to have a whole half . . .'": Harmon, Amy, "Love You, K2a2a, Whoever You Are," *The New York Times*, January 22, 2006.

PP. 62–63 "Harvard University's Henry Louis Gates, Jr. . . .": Gates, Jr., Henry Louis, "My Yiddishe Mama," *The Wall Street Journal*, February 1, 2006, A16.

Chapter 5. Finding Aaron

P. 66 "He is what is called in Hebrew an *oleh* . . .": Telushkin, Joseph, *Jewish Literacy* (New York: William Morrow, 2001), 286.

PP. 66–67 "That area is controlled . . ." and entire Skorecki interview: Karl Skore-cki interview, August 7, 2002, and November 20, 2002.

P. 69 "Geneticists had already published a number of comprehensive studies . . .": Key studies of classical markers include Livshits, G. et al., "Genetic Affinities of Jewish Populations," *American Journal of Human Genetics* 49, 1991, 131–146; Karlin, S., Kenett, R., and Bonné-Tamir, B., "Analysis of Biochemical Genetic Data on Jewish Populations: I & II," *American Journal of Human Genetics* 31, May 1979, 324–365; Carmelli, D. and Cavalli-Sforza, L.L., "The Genetic Origin of the Jews: A Multivariate Approach," *Human Biology* 51, March 1979, 41–61.

PP. 71–72 "The Aaronites lost their priestly status . . .": Spencer, John R., "Aaron" in *The Anchor Bible Dictionary* Vol. I (New York: Doubleday, 1992), 1–6; Rehm, Merlin D., "Levites and Priests" in *The Anchor Bible Dictionary* Vol. IV (New York: Doubleday, 1992), 297–310.

P. 72 "Cohanim became subject to special laws . . .": Kleiman, Ya'akov, "Who Is a Cohen? The Tribe—The Cohen-Levi Family Heritage," http://www.cohen-levi.org/02whois.htm.

P. 74 "her studies of the ancestry of Jewish and Arab ethnic populations . . .": See Bonné-Tamir, B. et al., "Polymorphism in Israel: An Overall Comparative Analy-sis," *Tissue Antigens* 2, 1978, 235–250; Bonné-Tamir, B., "Oriental Jewish Com-munities and Their Genetic Relationship with South-West Asian Populations" in *Genetic Microdifferentiation in Human and Other Animal Populations,* edited by Y.R. Ahuja and J.V. Neel (Delhi, India: Indian Anthropological Association, 1985), 153–170.

P. 74 "'I was immediately taken by Karl's idea . . .'": Batsheva Bonné-Tamir inter-view, November 19, 2002.

P. 75 "To have Karl raise this issue . . .": Michael Hammer interview, August 15, 2002.

P. 77 "I'm a park-the-car-around-the-corner attendee of an Orthodox shul . . .": Neil Bradman interview, December 3, 2002.

P. 78 "Stocked with what looked like tubes of mouthwash . . .": Description from Hirschberg, Peter, "Decoding the Priesthood," *The Jerusalem Report*, May 10, 1999, 30–35.

P. 79 "'This may have been the founding modal haplotype . . .'": Skorecki, Karl et al., "Y Chromosomes of Jewish Priests," *Nature* 385, January 2, 1997, 32–33.

P. 79 "'It was incredibly exciting . . .'": Hirschberg, Peter, "Decoding the Priest-hood," *The Jerusalem Report*, May 10, 1999, 32.

P. 80 "'It's emotionally very charged . . .'": Grady, Denise, "Father Doesn't Al-ways Know Best," *The New York Times*, January 19, 1997, Section 4, 4.

P. 81 "'The simplest, most straightforward explanation . . .'": Grady, Denise, "Finding Genetic Traces of Jewish Priesthood," *The New York Times*, January 7, 1997, 6.

P. 81 "'The Priests' Chromosome?'": Travis, J., "The Priests' Chromosome? DNA Analysis Supports the Biblical Story of the Jewish Priesthood," *Science News* 154, October 3, 1998, 218.

P. 81 "The Jewish Telegraphic Agency . . .": Cohen, Debra Nussbaum, "Kohen Gene Pioneers Fear Misuse," *Jewish Telegraphic Agency*, January 7, 1997.

P. 82 "Until that time, few Arabs lived in the Holy Land . . .": Taylor, Jane, "Masters of the Desert," *Eretz Magazine*, March–April 1995, 43–56.

P. 84 "Jewish visitors in the eighteenth and nineteenth centuries . . .": Roman, Yadin, "Pilgrimmage to Aaron's Tomb," *Eretz Magazine*, March–April 1995, 19–26.

P. 85 "'We realized that more genetic markers . . .'": Travis, John, "The Priests' Chromosome? DNA Analysis Supports the Biblical Story of the Jewish Priesthood," *Science News* 154, October 1998, 218.

PP. 85–86 "'With their claims of common ancestry . . .'": Mark Thomas interview, December 3, 2002.

P. 86 "Goldstein appreciated the potential of using microsatellites . . .": For a review of the uses of Y-chromosomal analysis, see Goldstein, David B. and Chikhi, Lounès, "Human Migrations and Population Structure: What We Know and Why It Matters," *Annual Review of the Genomics of Human Genetics* 3, 2002, 129–152.

P. 86–87 "'I recall getting an e-mail . . .'": David Goldstein interview, December 3, 2002.

P. 87 "'What we found was that approximately 50 percent of the Cohanim . . .'": Karl Skorecki e-mail to author, February 29, 2004.

P. 88 "one of the streamlined grouping of eighteen major lineages . . .": Hammer, Michael et al., "A Nomenclature System for the Tree of Human Y-Chromosomal Binary Haplogroups," *Genome Research* 12, 2002, 339–348.

P. 88 "Subgroups, or *clades*, of two haplogroups . . .": Nebel, Almut et al., "The Y Chromosome Pool of Jews as Part of the Genetic Landscape of the Middle East," *American Journal of Human Genetics* 69, November 2001, 1095–1112; Quintana-Murci L. et al., "Y-chromosome Lineages Trace Diffusion of People and Languages in Southwestern Asia," *American Journal of Human Genetics* 68, October 2001, 537–542.

P. 88 "It is found at its highest variety . . .": Behar, Doron et al., "Contrasting Patterns of Y Chromosome Variation in Ashkenazi Jewish and Host Non-Jewish European Populations," *Human Genetics* 114, 2004, 354–365.

P. 89 "the person who carried the haplotype . . .": Thomas, Mark G. et al., "A Genetic Date for the Origins of Old Testament Priests," *Nature* 394, July 9, 1998, 138–140.

P. 90 "Although the founding . . .": Porter, Stanley R., "Zadok" in *The Anchor Bible Dictionary* Vol. VI (New York: Doubleday, 1992), 1034–1036.

P. 91 "Bradman and Thomas struck a similar tone . . .": Bradman, Neil and Thomas, Mark, "Genetics: The Pursuit of Jewish History By Other Means," *Judaism Today*, Autumn 1998, 6.

PP. 91–92 "'The suggestion that the "Cohen Modal Haplotype" . . .'": Zoossmann-Diskin, Avshalom, "Are Today's Jewish Priests Descended from the Old Ones?" *HOMO: Journal of Comparative Human Biology—Zeitschrift fuer vergleichende Biologie des Menschen* 51, 2001, 156–162.

P. 91 "Iraqi Kurds": Brinkmann, C. et al., "Human Y-chromosomal STR Haplotypes in a Kurdish Population Sample," *International Journal of Legal Medicine* 112, 1999, 181–183.

P. 91 "Armenians": Correspondence between Dr. Levon Yepiskoposyan (Yerevan, Armenia), head of the Institute of Man and president of the Armenian Anthropological Society, with Kevin Brook, cited in Brook, Kevin A., "The Origins of East European Jews," *Russian History/Histoire Russe* 30, Spring–Summer 2003, 1–22.

P. 91 "southern and central Italians": Caglià, A. et al., "Increased Forensic Efficiency of a STR-based Y-Specific Haplotype by Addition of the Highly Polymorphic DYS385 Locus," *International Journal of Legal Medicine* 111, 1998, 142–146.

P. 91 "Hungarians in the Budapest region": Woller, Füredi, J. et al., "Y-STR Haplotyping in Two Hungarian Populations," *International Journal of Legal Medicine* 113, 1999, 38–42.

P. 91 "Palestinian Arabs": Nebel, Almut et al., "The Y Chromosome Pool of Jews as Part of the Genetic Landscape of the Middle East," *American Journal of Human Genetics* 69:5 (November 2001), 1095–1112.

PP. 91–92 "'It's not a coincidence that the haplotype was named the "Cohen Modal Haplotype"'": Avshalom Zoossmann-Diskin interview, November 2002.

P. 92 "In most scholarly contexts . . .": Marks, Jonathan, *What It Means to Be 9% Chimpanzee* (Berkeley: University of California Press, 2002), 247.

P. 94 "two small silver scrolls inscribed with a Hebrew prayer . . .": Kleiman, R. Ya'akov, "DNA and the Chain of Tradition," *Yated Ne'Eman*, July 7, 2000, 9–12.

P. 94 "This is powerful . . .": Hirschberg, Peter, "Decoding the Priesthood," *The Jerusalem Report*, May 10, 1999, 34–35.

P. 94 "Jacob grew up disaffected, 'a bagels and lox kind of Jew'": Ya'akov Kleiman interview, November 26, 2002.

P. 95 "the Bible says Jesus was not of his seed . . .": For one cogent analysis of the nativity narratives, see Callahan, T., *Secret Origins of the Bible* (Altadena, CA: Millennium Press, 2002), 374–404.

P. 96 "Davidic lineage can be traced from . . .": Einsiedler, David, "Are You a Descendant of King David?" *Rabbinic Genealogy Special Interest Group Online Journal*, http://www.jewishgen.org/Rabbinic/journal/kdavid.htm.

P. 96 "Susan Roth . . .": Roth, Susan, http://www.davidicdynasty.org.

Chapter 6. The Crystallization of Jewishness

P. 100 "The full five books did not take final shape . . .": Thompson, Thomas L., *The Mythic Past: Biblical Archaeology and the Myth of Israel* (New York: Basic Books, 1999); Lemche, Niels, "Early Israel Revisited," *Currents in Research: Biblical Studies* 4, 1996, 9–34; Davies, Philip R., "In Search of 'Ancient Israel,'" *Journal for the Study of the Old Testament Supplement Series—JSOTS* 148 (Sheffield, UK: Sheffield Academic Press, 1995). For a critique of this claim, see Dever, William G., *What Did the Biblical Writers Know and When Did They Know It?* (Grand Rapids, MI: Wm. B. Eerdmans Publishing, 2001).

P. 100 "the Israelite kingdom arose out of a core group of Canaanites . . ." Dever, William G., *Who Were the Israelites and Where Did They Come From?* (Grand Rapids, MI: Wm. B. Eerdmans Publishing, 2003), 219–220.

P. 101 "The emergence of Israel . . .": Finkelstein, Israel and Silberman, Neil Asher, *The Bible Unearthed* (New York: Free Press, 2001), 48–71.

PP. 102–103 "The Bible claims . . .": Machinist, Peter, "Administration of (Assyro-Babylonian) Palestine," *The Anchor Bible Dictionary* Vol. V (New York: Doubleday, 1992), 69–81.

P. 104 "Rather than a Davidic golden age . . .": Finkelstein, Israel and Silberman, Neil Asher, *The Bible Unearthed* (New York: Free Press, 2001), 229–250.

P. 105 "The great prophets . . .": Marcus, A.D., *Rewriting the Bible* (New York: Little Brown, 2000), 46–47.

P. 106 "the victorious Babylonians deported . . .": Finkelstein, Israel and Silberman, Neil Asher, *The Bible Unearthed* (New York: Free Press, 2001), 296–313.

P. 107 "evidence of the Israelites as merchants . . .": Barnavi, Eli (ed), "Babylon and Egyptian Jewry: The Person Era," *A Historical Atlas of the Jewish People* (New York: Schocken, 1992), 30–31.

P. 107 "Judea was little more . . .": Williamson, H.G.M., "Administration of (Judean Officials) Palestine," *The Anchor Bible Dictionary* Vol. V (New York: Doubleday, 1992), 81–86.

P. 108 "Some rabbinical authorities . . .": North, Robert, "Ezra," *The Anchor Bible Dictionary* Vol. II (New York: Doubleday, 1992), 726–728.

P. 109 "establishing what amounts to . . .": North, Robert, "Administration of (Judean Officials) Palestine," *The Anchor Bible Dictionary* Vol. V (New York: Doubleday, 1992), 86–90.

P. 109 "Although the zealous Ezra and Nehemiah . . .": Cohen, Shaye J.D., *The Beginnings of Jewishness* (Berkeley: University of California Press, 1999), 263–273, 308–340.

P. 111 "Traditional Jewry maintained . . .": Ben-Zvi, Itzhak, *The Exiled and the Redeemed* (Jerusalem: The Jewish Publication Society of America, 1976), 123–130.

PP. 111–112 "From a nation of hundreds of thousands . . .": Historical information on the Samaritans verified in part from an e-mail from Benyamim Tsedaka, received March 4, 2004.

P. 112 "Historians have long been skeptical . . .": Talmon, Shemaryahu, "Biblical Traditions in Samaritan History" in Stern E., Eshel H. (eds), *The Samaritans* (Jerusalem: Yad Ben-Zvi Press, 2002), 692.

PP. 112–113 "Bonné-Tamir has found . . .": Bonné-Tamir, B., "Maternal and Paternal Lineages of the Samaritan Isolate: Mutation Rates and Time to Most Recent Common Male Ancestor," *Annals of Human Genetics* 67 (Pt. 2), March 2003, 153–164.

P. 113 "The scientists speculate . . .": Shen, Peidong et al., "Reconstruction of Patrilineages and Matrilineages of Samaritans and Other Israeli Populations from the Y-Chromosome and Mitochondrial DNA Sequence Variation," *Human Mutation* 24, 2004, 248–260.

PP. 113–114 "After two centuries of Persian rule . . .": Fischer, Thomas, "Administration of (Ptolemaic/Seleucid) Palestine," *The Anchor Bible Dictionary* Vol. V (New York: Doubleday, 1992), 90–96.

P. 114 "Jewish diaspora outposts . . .": Scheindlin, Raymond P., *A Short History of the Jewish People* (New York: Oxford University Press, 1998), 25–49.

PP. 114–115 "The Hellenized Jewish intelligentsia embraced . . .": Johnson, Paul, *A History of the Jews* (New York: Harper, 1987), 81–168.

P. 119 "The Apostles stress a new covenant . . .": For an informative discussion of the different evolutionary strategies of ancient Jews and Christians, read Williams, Patricia A., "The Fifth R: Jesus as Evolutionary Psychologist," *Theology and Science* 3, July 2005.

P. 120 "before and after the time of Jesus . . .": Barnavi, Eli (ed), "Between Jerusalem and Alexandria," *A Historical Atlas of the Jewish People* (New York: Schocken, 1992), 36–37.

P. 121 "deported to Italy": Friedenberg, Daniel M., "Early Jewish History in Italy," *Judaism*, Winter 2000.

P. 122 "the Seleucid ruler in Syria . . .": Johnson, Paul, *A History of the Jews* (New York: Harper, 1987), 134.

P. 124 " 'Claudia Aster . . .' ": Brooks, Andrée, "In Italian Dust, Signs of a Past Jewish Life," *The New York Times*, May 15, 2003.

PP. 124–125 "Beleaguered Palestinian Jews . . .": Friedenberg, Daniel M., "Early Jewish History in Italy," *Judaism*, Winter 2000.

P. 127 "The new orthodoxy . . .": For a provocative analysis of this critical moment in Jewish-Christian relations, see Carroll, James, *Constantine's Sword: The Church and the Jews* (Boston: Houghton Mifflin, 2001).

P. 127 "Jews were also forbidden . . .": Barnavi, Eli (ed), "When the Roman